Real World Scanning and Halftones

REAL WORLD SCANNING AND HALFTONES

The definitive guide to scanning and halftones from the desktop

BY DAVID BLATNER
AND STEVE ROTH

An Open House Book

Peachpit Press

Real World Scanning and Halftones
David Blatner and Stephen F. Roth

Peachpit Press, Inc.
2414 Sixth St.
Berkeley, California 94710
800/283-9444
510/548-4393 (phone)
510/548-5991 (fax)

Permissions

Permissions and other copyright notices appear on page 273.

Trademarks

Notice of Liability

ISBN 1-56609-093-8
Printed and bound in the United States of America
0 9 8 7 6 5 4 3 2 1

PREFACE

PRIMAL SCREENS

It was 6:30 PM on July 28, 1987. Two hours earlier, Steve had handed in the manuscript for *Real World PostScript.*

Isn't it always the case that as soon as you create something, you think of a half dozen ways it could have been better? That's what happened to Steve that day. He started thinking of sections he could have added, ways he could have made it more accessible to non-technical readers, better ways of illustrating concepts, and half a dozen other tweaks and corrections he wanted to incorporate.

He was loaded with ideas, but hey—he'd just turned in a whole book, and he wasn't going to go back and do it all again. As it turned out, more than five years went by before he got around to it.

AS TIME GOES BY

In the years since *Real World PostScript* was published, desktop scanning and halftoning has changed immensely. The hardware, the software, and the body of knowledge necessary to produce top-notch desktop-published documents—in black-and-white, grayscale, and color—have evolved to the point that desktop publishers now have the tools to equal, and in many cases surpass, "traditional" techniques. The tools *finally work* now, which is more than we can say for the early days.

Another important thing happened in those years. Steve met David Blatner. David was pushing PostScript and desktop publishing to the limits, doing what in those days "simply couldn't be done."

When we worked together on *ScanJet Unlimited*, back in 1989, we both realized that the world needed another book—one that included the technical background of *Real World PostScript,* and had the practical, step-by-step advice of *ScanJet Unlimited*, but that was applicable to any scanner, any computer, and any desktop publisher.

So we spent years amassing all the information we could on desktop scanning and halftones, figuring someday we'd finally get around to writing this book. Whether Steve was writing yet another scanner roundup for *Macworld*, or David was writing yet another chapter for *The QuarkXPress Book*, it was an opportunity to scour the literature, plumb the online services, grill the experts, and generally pull together as many tips, tricks, and techniques as we could on the subject.

PULLING IT TOGETHER

We were busy, though. We wrote five books, launched a conference company, and lectured to thousands of people around the country.

It was obvious from talking to all those desktop publishers that people really wanted the information we'd gathered, and that the information was simply unavailable from any other source. So, more than five years after those first glimmerings, we finally decided to do it—pull together the big pile of information we'd been gathering (there's several cubic feet of literature in a box in David's office), and put out the book we always wished someone else would write, so we didn't have to. The result is in your hands.

REAL WORLD SCANNING AND HALFTONES

The main thing we wanted to do with this book was provide step-by-step, type-in-the-numbers instructions for getting great-quality images out of your scanners, laser printers, and imagesetters, using a variety of software. And we've done that.

But there's simply no way we can provide you with the exact settings you need for every piece of hardware and software in every situation you'll find yourself in. So we've also done our best to give you the conceptual underpinnings you need to address your own publishing environment.

Because we want you to understand not only what button to push, but also why you should push it (or not), you'll find some lengthy conceptual explanations here. Hopefully we've presented them so that you can grasp them easily and without the pain that we've endured figuring this stuff out for ourselves.

Finally, while there are plenty of books out there on imaging theory, hundreds on printing, and thousands of application-specific books, almost nothing is informed by a real-world understanding of the whole production process—from scanning to imagesetting to offset printing. What none of these books brings to the party is the understanding of how all these pieces interrelate. And if there's one thing we've learned in the world of scanning and halftones, *everything* relates to everything else.

How to Read this Book

So how should you read this book? Since we wrote it, we think you should read it cover-to-cover. In a single sitting. Twice.

We realize that's a bit unrealistic. Because all of this stuff is so interrelated, though, it's hard to give any other advice. In some ways, the topics in this book are like Steve's favorite obsession: sailboat racing. You just can't understand sail shape without a good knowledge of hull dynamics, or vice versa. Similarly, you can't understand scanning techniques if you don't know about printing presses, and so on.

There's no single chapter in this book that doesn't rely on information in some other chapter (or several other chapters). Everything depends on everything else. So when you read the book the first time, some things may appear somewhat mysterious. But when you read it again, the patterns and relationships will start to emerge, and techniques that at first appeared odd will begin to make more sense.

FIRST STEPS

Nonetheless, if you just want to dip into this book quickly, we suggest you begin with Chapters 1, 2, and 3, then jump to Chapters 19, 20, and 21. They'll give you a basic grounding in the essentials of scanning and halftones. From there you can wander through the other chapters, filling in the nooks and crannies of your expertise.

We Had Help

There's no way that any two people could have enough personal experience to address all the diverse areas that are encompassed in desktop scanning and halftones. So, while we'd love to take credit for the entire book project, we had help. Lots of it.

Here's a short list of the people who helped us create this book. What you find of value here probably came from one of them. What you find useless or just plain superfluous . . . well, that part we came up with ourselves.

Efi Arazi of EFI
Tom Bernard of Bersearch
Henry Bortman of *MacUser*
Frank Braswell of Systems of Merritt
Luis Camus of *Macworld*
Cary Cartmill of Pacific Color
Rob Cook of Light Source
John Cornicello of Thunder Lizard Productions
Peter Fink of *Desktop to Press*
Bruce Fraser of *Macweek* and *MacUser*
Harvey Grosser
Jim Hamilton of Linotype-Hell
John Knoll
Paul McAfee and Joe Runde of Kodak
Herb Paynter of Image Express
Doug Peltonen of Aldus
Phil Rose of Varityper
Margo Schulter

Joanna Sinwell of Hewlett-Packard
Steve Stankiewicz
Simon Tuckett
Greg Vander Houwen of Interact and Ivey Seright
Steve "dreams in PostScript" Werner
Pat Wood
"Wildman" Bill Woodruff, formerly of Adobe Systems.

... AND A CAST OF THOUSANDS

Behind every great author stands a myriad of great supporters. We were very lucky to have a number of people who also helped in essential ways. Glenn Fleishman at Open House wrote, edited, and brainstormed over many parts of the book. Steve "It doesn't matter! They all look the same!" Broback and Marci Eversole of Thunder Lizard Productions, and Don Sellers of Stourwater Productions provided invaluable encouragement. Olav Martin Kvern, master of shapes and sounds, was always there as a sounding board.

FROM STEVE: "Thanks to Susie, Jesse, and Dia for doing without me so much while we worked on this book. If I hadn't known that I would be coming home each night, leaving each morning would have been much harder."

FROM DAVID: "My sincere thanks goes to all my friends who suffered through months of my jabbering about the obscurities of digital imaging. My housemates, Ann, Marisa, and Jasmine. And, my mother, without whom I would probably forget to wear a sweater."

Overview

Part 1 Halftones

Part 2 Scanning

Part 3 PostScript

Part 4 Applications

Contents

PART IV: APPLICATIONS

PART 1
HALFTONES

1 Dots, Spots, and Halftones

How Halftones Work

It's Picasso's worst nightmare. Imagine a world that knows no color. No reds, no greens, no blues; in this world, there aren't even any grays. In this world, the sky is white, the ground is white, the horizon is white. The only color that breaks the monotony of endless white is black. Little spots of black.

No, this isn't some weird scenario from *The Twilight Zone.* It's the way that graphic artists and lithographers have worked for the past hundred years. Offset, letterpress, and gravure printing presses can't reproduce colors or any level of gray without changing inks. The same thing goes for dot-matrix and laser printers. In other words, there's no way to print a continuous-tone photograph (one that contains a bunch of grays), without running the paper through the printer, then changing the ink to a different shade of gray, then running the paper through again, then changing the ink, then . . . but, wait. That *can't* be how it's done; it's too complicated.

And, in fact, it's not.

The Dot's the Thing

Lithographers figured out in the late 19th century that they could create a tint of a colored ink by breaking the color down into a whole bunch of little spots. They could make gray, for instance, by printing black spots really close to each other. Our eyes and brain help by "lying" to us, telling us we see gray instead of black spots.

This process of breaking a gray image up into black spots is called halftoning. Figure 1-1 shows an everyday halftone of a photograph with a blown-up section showing the halftone spots.

FIGURE 1-1
A halftoned photograph

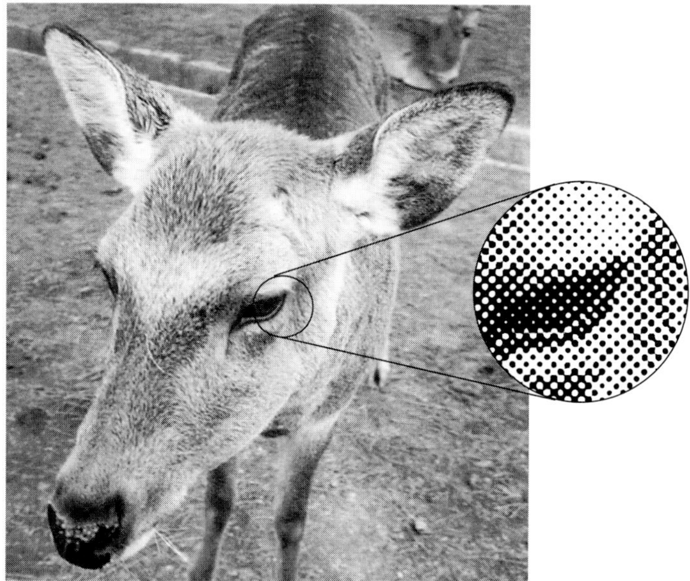

PHOTOGRAPHIC HALFTONES

Traditionally, halftones are photomechanical reproductions. Someone puts a piece of photosensitive film behind a finely etched *screen* (originally made of glass, but now usually of film)—and projects light through the photograph, which exposes the film (see Figure 1-2). The result is a pattern of evenly spaced spots of different sizes. Because of *diffraction*—the tendency of light to spread after passing through a narrow aperture—the spots in dark areas are big (even overlapping), and the dots in light areas are small.

TINTS AND TINT PERCENTAGES

Bear in mind that halftoning isn't just for photographs. Any time you see gray printed on a page, it's the result of halftoning. Flat tint areas

FIGURE 1-2
Photographic halftoning

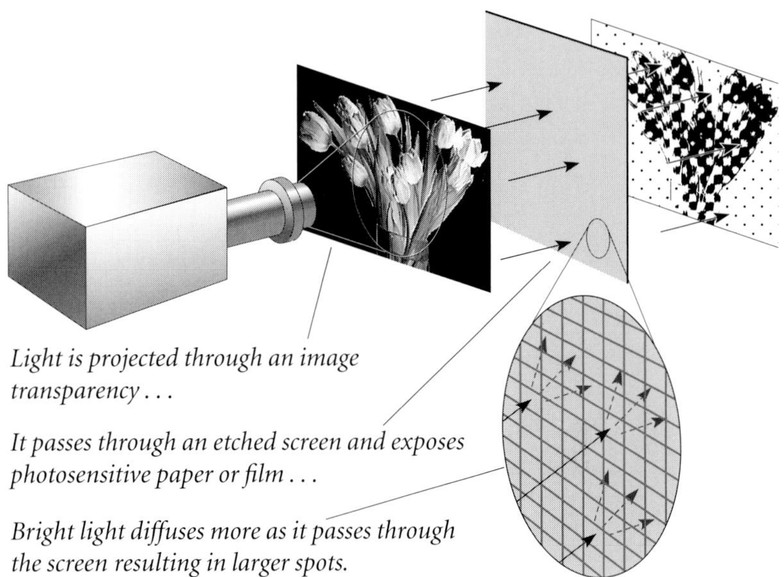

are made using the same method described above, except that light is projected directly through a screen rather than passing through a photographic image first.

Tints are referred to in percentages. A light gray tint, where the halftone spots are a tenth of their full size, is a 10-percent tint. A medium gray area might be a 50-percent tint, and in a dark gray area—let's say a 90-percent tint—the spots are 90 percent of full size. A 100-percent tint, of course, is black (see Figure 1-3).

FIGURE 1-3
Tint percentages

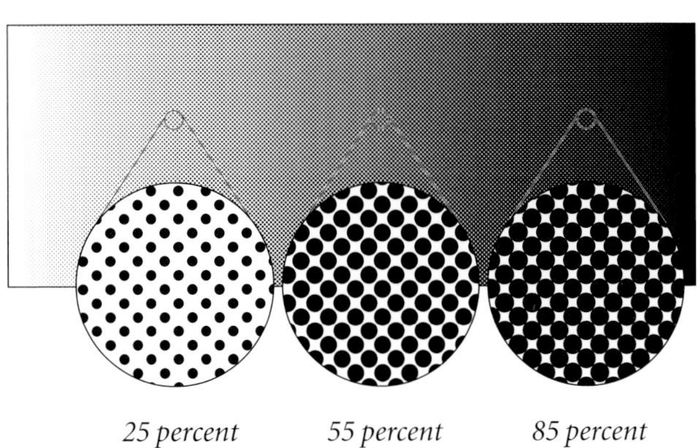

Digital Halftones

These days, photographic halftones are less and less common. Computers, laser printers, and imagesetters are the main tools for creating halftones—though, of course, they don't use screens made of glass or plastic to make them. In order to understand how computers make halftones, we need to stop for a moment and look at how these devices make images in the first place.

Dots

Let's face it: computers are really stupid. In fact, they're so dumb that they only understand two things: zero and one. You can think of this as off and on, or—even better—white and black. When a laser printer is creating an image, it simply turns the laser on and off—on where the paper should be black, and off where it should be white. The dots that the laser creates are so small and close together that two or more dots in a row appear as an area of solid black.

The dots that most desktop laser printers create are tiny. In fact, you can fit three hundred of them next to each other in a single inch. That's why people call them 300-dpi (dot-per-inch) printers. Other printers (typically plain paper output) and imagesetters (photosensitive paper and film) can create even smaller dots, as fine as 1,200, 2,400, or even 3,000 dpi—dots which are often too small for the human eye to see.

Spots

But how can you make halftone spots larger and smaller when a laser printer's dots are all the same size? The solution is pretty simple. The computer groups together a bunch of dots into a single halftone cell. This cell is a square grid of dots, each of which can be turned on or off. To create a dark area (a large spot), lots of the dots are turned on; to create a light area (a small spot), only a few dots are turned on (see Figure 1-4).

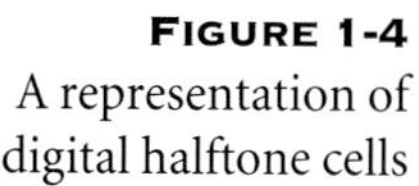

FIGURE 1-4
A representation of digital halftone cells

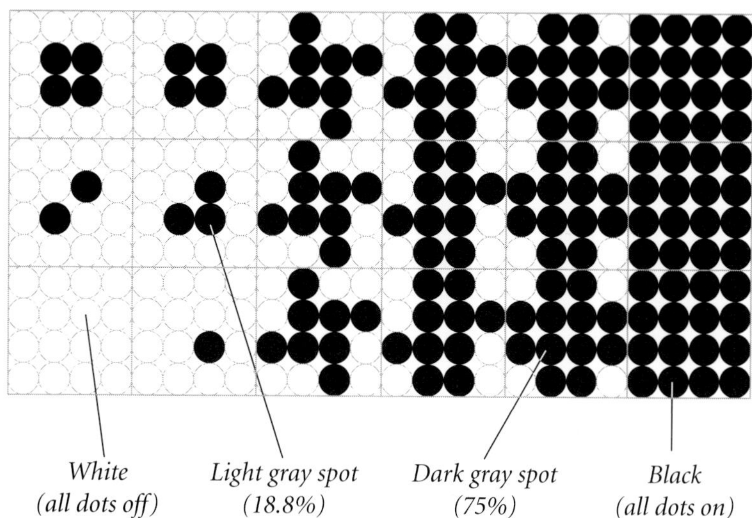

The important concept here is that the spacing of the halftone spots doesn't change. They're not closer together in dark areas. Only the number of dots turned on *within each cell* changes. The group of dots that are turned on is the halftone spot (see Figure 1-5). We'll talk more about these spots and their shapes in the next chapter.

FIGURE 1-5
Halftone spots

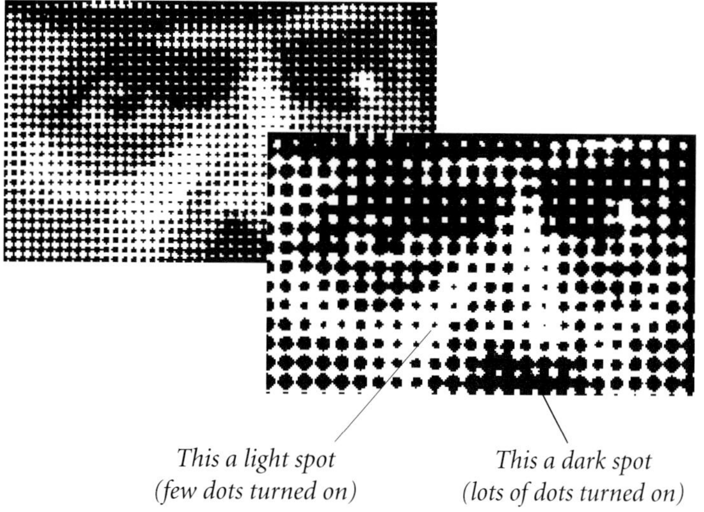

Halftones, Tints, and Screens

There's one other piece of terminology that we want to clarify before we end this chapter—*screening*. We mentioned above that the etched piece of glass or film that's used to create photographic halftones is called a screen. The process of creating the halftone pattern is called screening. To confuse matters more, the pattern of dots that results from screening—the halftone pattern—is also called a screen.

The more you recognize halftones and how they're used and created, the more effective you can be in your own work. Next we're going to look at some more details about halftoning: frequency, angle, and spot shape.

2 FREQUENCY, ANGLE, AND SPOT SHAPE

WHAT MAKES A HALFTONE

In this chapter we talk about one of David's favorite things: patterns. David goes ga-ga over patterns: intricate embroidery, vaulted Islamic ceilings, the repeating natural fractal patterns of age-old trees . . . and, of course, halftones.

As we noted in the last chapter, halftones are made up of bunches of spots, some lighter and some darker. Each spot is in a cell made up of printer dots. These halftones cells are arranged in a pattern—a pattern that has three primary attributes: frequency, angle, and spot shape. Let's look at each of these in turn.

SCREEN FREQUENCY

Imagine eggs in an enormous egg carton. Each egg sits in its own place, with an equal distance to the egg on the left and the egg on the right. If the egg carton were big enough, you wouldn't say it was a six-egg carton or a 12-egg carton. Rather, you'd say there were 12 eggs per foot or 39 eggs per meter, or something like that. Each row is the same distance from the next row, so you can say that the "egg-per-foot" value (epf) is the number of rows of eggs per foot (see Figure 2-1).

Now let's look back at halftones: you've got a whole bunch of spots (eggs) that sit in rows of a big grid (egg carton). There are too

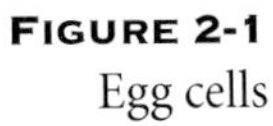

FIGURE 2-1
Egg cells

many spots to count, so you simplify the problem by counting the number of spots—or rows of spots—per inch. This is called the frequency of the halftone, because it tells you how frequently spots occur within an inch. The frequency is typically measured in lines per inch (lpi). Of course, in metric countries, it's lines per centimeter (L/cm).

Specifying screen frequency in lines per inch confuses the issue slightly, because we're not really talking about lines here. We're actually talking about rows of spots. But if you keep that in mind you won't go wrong.

People often talk about coarse and fine screens. The lower the screen frequency, the coarser the screen. That is, the cells are bigger and the image is rougher looking. The higher the screen frequency, the finer the image, because the cells are really small (see Figure 2-2).

FIGURE 2-2
Screen frequency

Screen Angle

If we go back to our egg story ("An egg-cellent idea," says Steve), we see that the pattern of all those eggs not only has a frequency, but an angle, too. If you look at the eggs straight on, the rows of eggs are at an angle of 0 degrees. If you turn the carton, though, the angle changes. Turn it to "one o'clock" and the angle is at 30 degrees. Turn it more, and the angle goes to 45 degrees (see Figure 2-3).

Figure 2-3 Turning the egg carton

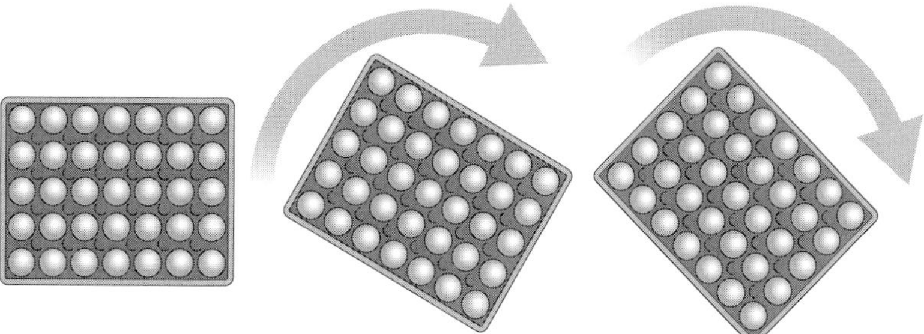

You can rotate a halftone screen almost as easily as turning an egg carton, and the angle to which you align the rows of spots is called the screen angle. A standard screen angle is 45 degrees because it's the least noticeable—it does the best job of fooling our eye into seeing gray instead of a pattern of spots (see Figure 2-4). However, there are two good reasons to change the angle of a screen. First, for weird special effects. Second, for color separations (see Chapter 11, *Rosettes and Moirés*).

Figure 2-4 Halftone angles

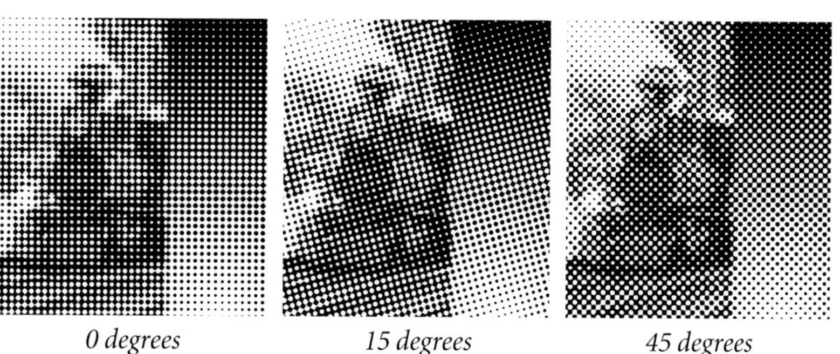

0 degrees *15 degrees* *45 degrees*

Spot Shape

The final element of halftone screens is the spot shape. Until now, we've been talking primarily about round spots. But remember that these spots are made of tiny dots, and we can arrange the dots any way we want. So, our spots can be circles, squares, triangles, lines, or even little pictures of Bart Simpson (see Figure 2-5).

Figure 2-5
Spot shape

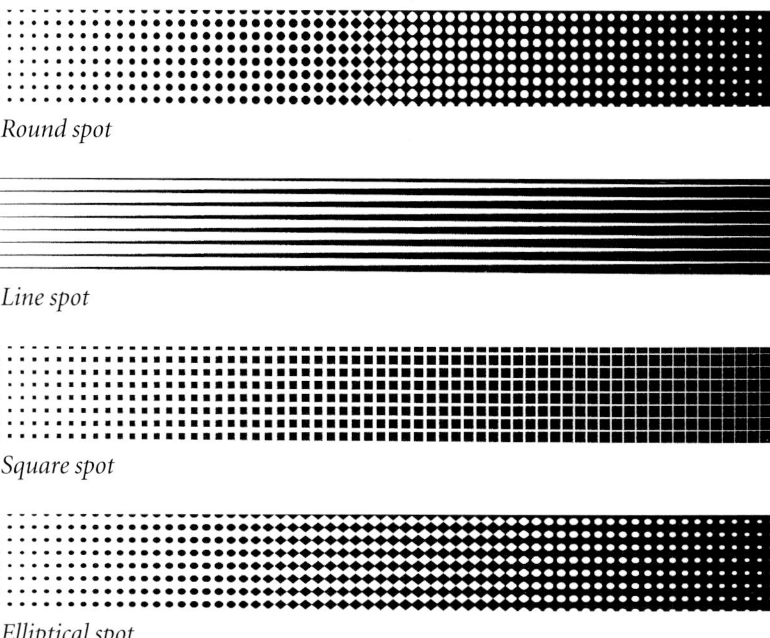

Round spot

Line spot

Square spot

Elliptical spot

You almost never notice the shape of a halftone spot unless the screen frequency is really low (in a 10-lpi halftone, there are 10 halftone cells per inch, so you can see the shape pretty easily; in a 133-lpi screen, each cell is so tiny—1/133 of an inch—that you can barely see what it looks like). However, the spot shape can make a difference in the appearance of your halftones. For example, a square spot often results in a sharper, higher-contrast look—especially in midtones—while an elliptical dot (actually a rounded-corner diamond) results in smooth transitions in graduated blends. Typically, though, you don't have to worry about spot shape unless you're trying to achieve a special effect.

There's more you should know about spot shapes, but we're going to put if off until Chapter 6, *The Glorious Spot.* However, we do think it's important to point out sooner rather than later that the egg-in-a-carton metaphor breaks down at this point. Do not (we repeat: do *not*) try to construct differently shaped eggs at home or in the office.

3 Frequency Versus Gray Levels

Tradeoffs in Digital Halftoning

Life is a series of compromises. We don't mean to get overly philosophical here, but it's true: in order to achieve one thing, you've got to give up something else. Well, it's no different with halftones. The essential trade-off in digital halftoning is between screen frequency and gray levels. Let's outline the problem first, and then take a little time to discuss why it's there and what you can do about it.

Screen Frequency, Gray Levels, and Output Resolution

Put succinctly, the halftone screen frequency has an inverse relationship to the number of gray levels possible at a given output resolution. Put more simply, the higher the screen frequency, the fewer levels of gray you can get at a given resolution. Reduce the screen frequency (or increase the output resolution), and you can get more levels of gray.

For example, on a typical desktop laser printer you can only get 10 different shades of gray (including black and white) in a 100-lpi halftone. If you make the halftone coarser—let's say 53 lpi, which is the default for most laser printers—you can get 33 different shades of gray.

POSTERIZATION

When you don't have enough gray levels, what results is *posterization*—an obvious stairstepping from one gray level to another, quite different, gray level—as opposed to a smooth transition. It's primarily a problem in graduated blends or halftones of photographs in which smooth transitions from white to black occur (see Figure 3-1). Posterization can also be an interesting special effect—but only you when you want it.

FIGURE 3-1
Posterization

SCARCITY OF DOTS

But why? Why should there be such a limitation on gray levels, when halftoning is so flexible? It's because the number of possible gray levels is determined by the number of dots in a halftone cell (see Figure 3-2). If a cell is made up of 25 dots, there are 26 possible gray levels (including white and black—all dots off and all dots on). If a cell is made up of 255 dots, there are 256 possible gray levels. The more dots per cell, the more possible gray levels.

The goal, then, is to work with halftone cells that have lots of dots in them, so you have lots of available gray levels. Therein lies the essential problem of digital halftoning—there are only so many printer dots to work with. The finer the halftone screen, the fewer

FIGURE 3-2
Dots per cell and gray levels

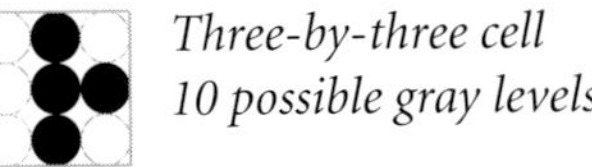

Three-by-three cell
10 possible gray levels

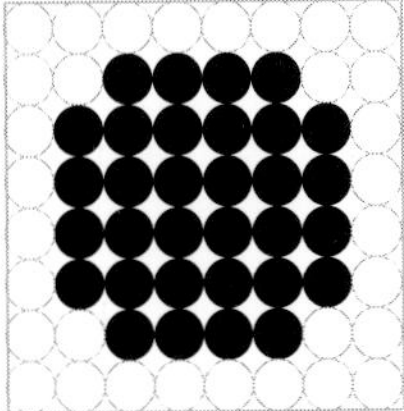

Eight-by-eight cell
65 possible gray levels

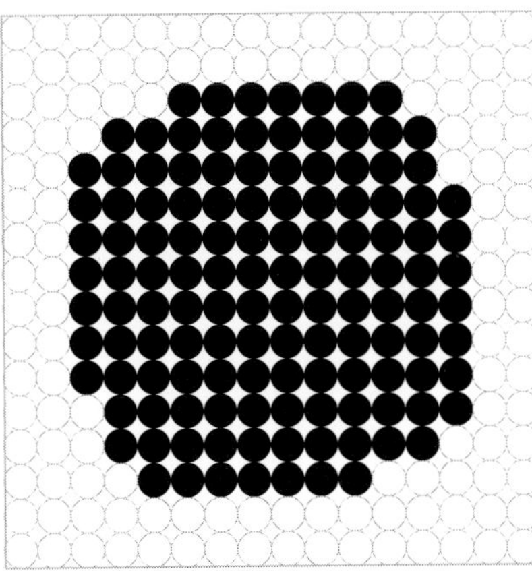

16-by-16 cell
257 possible gray levels

the number of dots are in each halftone cell, and therefore the fewer levels of gray you have available (see Figure 3-3).

You can determine the number of dots in a single halftone cell with simple arithmetic. (Don't worry, this book is written by two people who are so non-math-oriented that we use calculators to "go forth and multiply"; David has a degree in Theater Arts, and Steve in Theory of Literature.) If you divide the resolution of the printer by the screen frequency, you get the number of dots on one side of the halftone cell. Square this number (multiply it by itself) to find the total number of dots in the cell.

For example, if you divide 300 dpi (the resolution of the printer) by 75 (the screen frequency of the halftone), you get 4. Then multiply that by itself and you get 16. That's the number of dots that can possibly be turned on or off in each halftone cell. That means that

One inch
300 printer dots

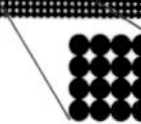

4-by-4 cell
17 gray levels
75-line screen

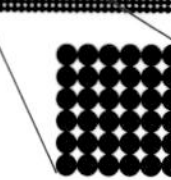

6-by-6 cell
37 gray levels
50-line screen

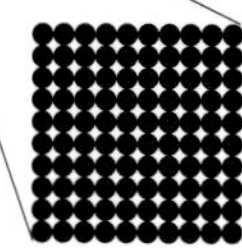

10-by-10 cell
101 gray levels
30-line screen

FIGURE 3-3 Halftoning at 300 dots per inch

when five dots are turned on in that cell, you have a 56-percent gray dot; when six dots are turned on, you get a 67-percent gray dot; and so on. You can't cut a dot in half, so there's no way to get any levels of gray in between. (Some laser printers can print dots of different sizes; see "Variable Laser Dots," below.)

To figure out the total number of levels of gray you can get, just add one to the number of dots in the halftone cell. In other words, if each level of gray is created by turning on an additional dot, you can get the same number of grays as there are dots in the cell . . . plus one. You have to add one if you want to include white as a level of gray (white is zero dots on). Figure 3-4 protrays the tradeoff graphically. For those of you who like formulas printed out, here it is:

$$\text{Gray levels} = (\text{output resolution} \div \text{screen frequency})^2 + 1$$

Here's an example: Let's say you're printing a halftone at 133 lpi on an imagesetter with a resolution of 2,400 dpi. Divide 2,400 by 133 and you get 18. Multiply this by itself and you get 324. Add one, and you see that you can get 325 levels of gray.

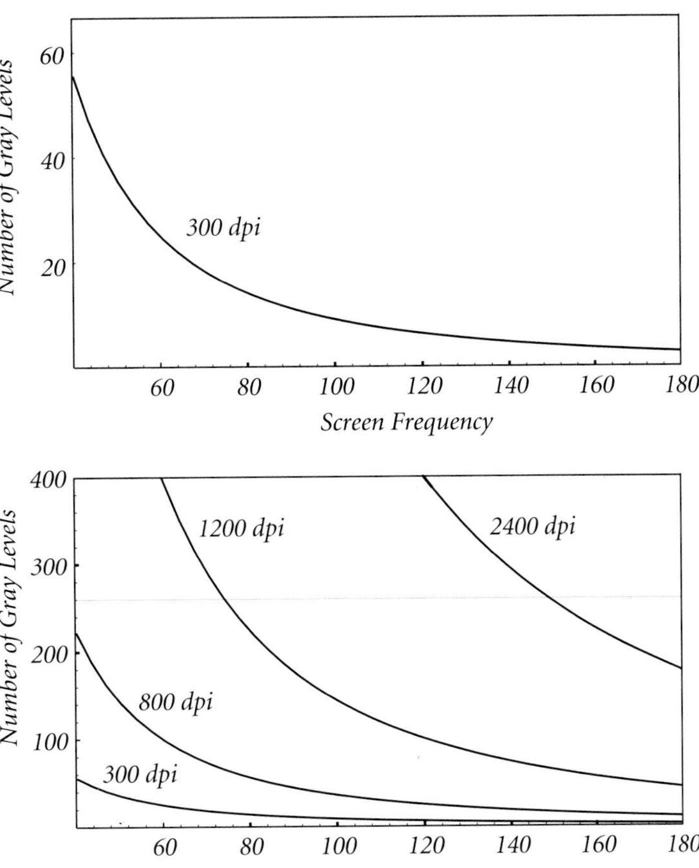

FIGURE 3-4
Gray levels versus screen frequency

Or can you? As it turns out, PostScript can really only create a maximum of 256 levels of gray. Anything over 256 levels of gray just gets thrown out.

There's a school of thought (propounded by Adobe Systems, the creator of PostScript, among others) that says the human eye can't perceive more than 256 levels of gray, but the research we've seen says that you can actually discern thousands of different gray shades. (We live in Seattle, where the weather is gray, cloudy, foggy, and misty so often that we've developed the ability to discern *and name* over 256 levels of gray.) What is certain is that printing presses typically can't reproduce more than 256 levels of gray.

Some experts say that a printing press can only produce about 100 levels of gray, so there's no need to strive for 256. That may be true, but if you have only 100 gray levels, and they're not apportioned properly over the whole-light-to-dark range, obvious posterization can result. For that reason, we suggest shooting to output the full 256 grays PostScript is capable of, thereby removing one bit of uncertainty in the whole process.

The Rule of Sixteen

PostScript's limitation of 256 levels of gray means that you rarely need to do the calculations described above. Instead, you can simply focus on two questions: What is the highest (finest) screen frequency you can use given the resolution of your printer? Or, what is the minimum resolution printer you can use given the screen frequency you're using?

The Rule of Sixteen gives you the answer to both these questions (it works because 16 is the square root of 256, but you don't really need to know that). You still have to do a little math, so don't put that calculator away too quickly. But this calculation is really easy. Here's the formula.

Maximum screen frequency = Output resolution ÷ 16
or
Required output resolution = Screen frequency × 16

For example, if you want to print a halftone at 150 lpi, multiply that by 16 to find that you need a minimum of 2,400 dpi to get 256 levels of gray. Or, if you know you'll be printing on a 1,200-dpi printer, you can divide that by 16 to find that you shouldn't print a halftone at higher than 75 lpi.

Variable Laser Dots

Many laser printers on the market these days can create variable-sized laser dots, which to some extent bypasses the essential tradeoff of digital halftoning, providing more gray levels at a given screen frequency and output resolution. Because the laser dots can vary in size, there is more control over halftone spot size, hence more potential gray levels. Apple's PhotoGrade technology, for instance, provides 91 gray levels with a 106-lpi screen frequency at 300 dpi.

Who Needs 256 Grays?

Many designers are now starting to realize that they really only need between 100 and 150 levels of gray in their scanned images (or even fewer). Graduated fills and some high-quality images almost always need 256 levels of gray (see Chapter 9, *Band Aid*), but if an image already has a lot of noise in it—or if you're printing on very absorbent paper—fewer gray levels might be fine for you.

By the way, if all your halftones are straight tints (no blends or photographs), then you probably don't have to worry about posterization and achieving a full 256 levels of gray. If you're producing a newsletter that has several 20-percent gray rectangles behind some type, you could easily print that at 120 lpi, even on a 1,200-dpi imagesetter. Sure, you can't get more than 101 shades of gray, but who cares? You've only got one shade of gray on your page!

4 Reproducing Halftones

Spot Variation and the Printing Process

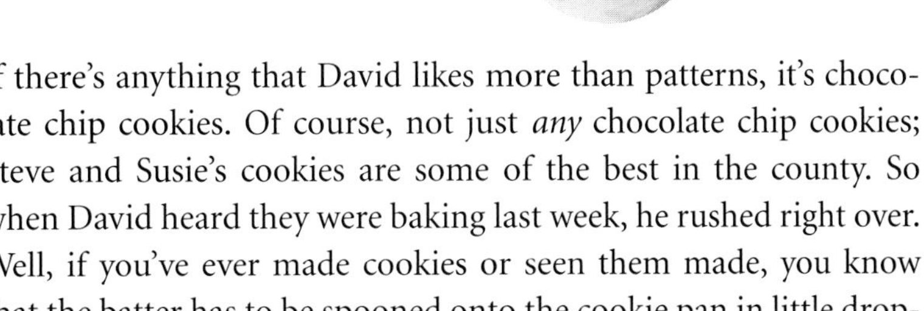

If there's anything that David likes more than patterns, it's chocolate chip cookies. Of course, not just *any* chocolate chip cookies; Steve and Susie's cookies are some of the best in the county. So when David heard they were baking last week, he rushed right over. Well, if you've ever made cookies or seen them made, you know that the batter has to be spooned onto the cookie pan in little droplets, and the droplets have to be a certain distance apart. Why? Because they melt and spread when you cook them.

On this particular day, Steve and Susie had put the cookie droplets a little too close to each other, and when David arrived all those glorious cookies had turned into one enormous blob. Of course Jesse and Dia, the Roths' adorable girls, were as happy as squirrels at a nut festival, and they dove hands-first into the mess.

Spot Variation

It turns out that cookies aren't the only thing that spread out and glob together: blobs of ink have the same properties. After all your hard work making a perfect halftone, you still have to contend with a printing press. Those thousands of tiny halftone spots spread out or shrink, depending on their size, the pressure of the offset cylinders, and (primarily) the absorbency and smoothness of the paper.

Sometimes they spread or shrink just a little, but sometimes they spread out so much that your image will look like the blob that ate Cincinnati (see Figure 4-1). As dark tints fill in and become black, your nice photograph ends up like a shot of Night Life in Sioux City.

The opposite problem also arises. The smallest spots tend to disappear (the tiny area of ink doesn't manage to adhere to the paper), so the subtlest tones in a snowbank blow out to white, or a 4-percent tint vanishes entirely.

The growth of halftone spots is traditionally called *dot gain.* This use of "dot" instead of "spot" is a holdover from conventional half-

FIGURE 4-1
Simulated spot variation on various paper stocks

Without spot variation

With spot variation. Note how shadows get darker (to the point of going black) and highlights get lighter (to the point of going white)

toning terminology, but we're going to call it *spot gain* to be consistent and avoid confusion. There's no real term for small spots disappearing, but we refer to the two problems together as *spot variation.*

Spot gain is measured in percentages: people talk about five- or 10-percent spot gain. Be aware, however, that five-percent dot gain doesn't mean that 50-percent spots become 55-percent spots. Spot gain percentages refer to the increase in area covered by a spot. So with 10-percent spot gain, a 20-percent spot goes to 22 percent, and 50 percent spot goes to 55 percent.

It's also important to understand that spot variation is nonlinear. All the spots in an image don't just increase by 10 percent on press (which is the impression given by the term *dot gain*). Small spots might get smaller, midtone spots might grow by a few percent, and spots in shadow areas might grow by 20 percent. If your software has a dialog box that lets you correct for dot gain by typing in a single percentage value, you should eye it with some suspicion.

Spot variation is more of a problem with fine screen frequencies printed on presses or plates that are unable to handle the small spot size, or on uncoated or textured paper. The spaces between spots are smaller at fine screen frequencies, so they clog up with ink quickly. In light areas, the spots are also small, so they're more likely to disappear entirely. If you work with a printer whose equipment and staff consistently reproduces screens of 175 lines per inch or higher, these problems are usually minimized or eliminated. For suggestions on setting screen frequency to control spot variation, see Chapter 5, *Setting Your Screens.*

What Causes Spot Variation?

Spot variation arises for a number of reasons throughout the process of turning digital information into real, physical spots on paper. Many factors can affect the reproduction of halftones; here's a list of a few of them.

- Laser density setting
- Density of imagesetter material

- Photographic reproduction
- Making plates
- Transfer of plate image to offset blankets, from blankets to impression cylinder, and from impression cylinder to paper
- Absorptive properties of paper, including texture
- Excessive inking
- Application of varnishes after inks

LASER DENSITY SETTING

If you're printing your final output from a laser printer, check to see if your laser printer has a density control (most Canon-engine printers, and many others, do). Turning this control to the lightest setting (for positive output) will often do a lot to improve the quality and reproducibility of halftones. It can also improve the quality of type and line-art graphics.

DENSITY OF IMAGESETTER MEDIA

Imagesetters expose photosensitive paper or film rather than applying toner to paper or transparencies. As a result, density is critically important to imagesetter output, just as it is to photographers developing their film. Let's look at some of the ways that density is affected in the imagesetting process.

Density. When imagesetter film or RC (resin-coated) paper is developed, the areas of emulsion that were exposed get transformed into another chemical compound, while the unexposed emulsion simply washes off. Density is a measure of how much of that chemical compound remains; essentially it's the compound's thickness or opacity.

You can measure this opacity using a densitometer, a device that either measures the amount of light that passes through film (transmission densitometry) or that is reflected back (reflection densitometry). A densitometer can also measure halftone tint percentages. Any service bureau running film must have a densitometer or else they can't calibrate their imagesetter.

Density is measured on a logarithmic scale, like the Richter Scale for earthquakes. A value of 4.0 means a 'quake (or density) 10 times greater than a value of 3.0. Film should have a density of roughly 3.5; paper should be around 1.7. Of course, these numbers vary according to material type, and they're almost never dead-on. In fact, density can usually be off as much as a full point in either direction for film, without causing problems. Consult your printers on ideal density for them, just as you would for ideal ink coverage or other printing variables.

If negative film being contacted onto plate material is not dense enough, highlights can be too dark, and shadow areas can clog up (spots tend to be enlarged). If the film is too dense, the highlights blow out and images can appear too light overall. The opposite is the case with positive film.

Imagesetter density controls. Imagesetters have density controls similar to laser printers. Just as you can over- or underexpose film in a camera—causing images to be too light or too dark—you can over- or underexpose the film or paper in the imagesetter.

Service bureaus often bump up the intensity of their imagesetters for paper output so that solid black areas such as type or line art are good and black. However, this causes halftone spots to get "bumped up" as well, becoming too large and potentially clogging up dark areas. A service bureau should have a dedicated machine for film, calibrated for that purpose.

The important rule here is to use a service bureau that calibrates its imagesetter. If you specify a 10-percent tint, you should get a 10-percent tint. A good test for a service bureau is to bring in a sample with tints, run it out on film or paper (depending on your dominant medium), and watch them check the tints on their densitometer. Percentages should be off by no more than two percent, and should generally be as close as one percent.

Chemical processing of imagesetter material. Also like camera film, imagesetter output is developed using chemical baths. Material goes through a developer bath, a fixative bath (broken out in photography as a stop bath and then a "fix" bath), and a wash. The developer and fix are heated, while the wash is at room temperature, and exists simply to squeegee and flush any remaining chemicals.

If the developer is below the correct temperature or depletes too fast, the density of the output is affected, because too much emulsion is washed off. If the fix is weak or depleted, film can be fogged, or paper may fade or yellow within a few days or weeks.

You should use a service bureau with a "replenishing-bath" processor. This type of processor introduces new chemicals at a constant rate, and maintains them at an even temperature. The wash water should either be replenished through a feed from a water line or through a frequently changed tank. You can tell if wash water is stagnant, as your output will have traces of algae on it.

Calibration. It's up to the service bureau to calibrate the whole system, from imagesetter through processor. The service bureau can run tests from one of their computers, to an imagesetter, through the processor, and finally check the results. With the correct software, a variety of tints and densities can be checked on the densitometer for the ideal value, and then any nonlinear values in tints can be corrected.

If your service bureau doesn't have a densitometer, or doesn't calibrate its equipment, it's impossible for you to trust the output. You could bring in a file on Monday and the same one on Tuesday, and have tint variations of 10 to 20 percent.

PHOTOGRAPHIC REPRODUCTION

Once you've pulled paper or film off the imagesetter or laser printer, it probably needs to be copied photographically in order to prepare plate-ready film. Any time you duplicate material photographically, you'll get some spot variation—as much as five- or 10-percent spot gain in dark areas for each generation of duplication—but more often in the two- to five-percent range.

The best solution is to avoid this step, using the film right off the imagesetter for burning plates. It's not always possible (there may be film stripping involved, for instance, and the printer might require a single piece of composite film for platemaking), but you can avoid spot variation if you can avoid duplicating film.

If for some reason the printer can't use the imagesetter film, there are some things you can do to reduce spot variation during photographic reproduction.

OUTPUT TO FILM, NOT PAPER. Shooting film from paper originals results in much more spot variation than shooting film from film. If you output film, the printer can make a "contact neg"—putting two pieces of film together and exposing one through the other. Contact negs generally result in one or at most two percent spot variation. If you supply paper, the printer has to actually take a picture of the paper, resulting in two- to four-percent spot variation.

Even if you're providing laser output, you can print on acetate (there are brands designed specifically for use with photocopiers and laser printers). This is especially helpful when you're using some of the high-resolution lasers and alternative halftoning methods, which are often more difficult to reproduce photographically (see Chapter 8, *Beyond the Spot*). If you *are* printing on paper from a laser printer, use a bright, hard, coated paper designed for laser printing of reproduction masters.

PROVIDE FILM THE WAY THE PRINTER NEEDS IT. You can reduce the number of generations of photographic reproduction that your material needs to go through by providing film in the proper format. If you provide positive film and the printer needs negative, he may need to add a duplication step (producing an *interneg*). Ask the printer if he or she wants emulsion up or down, negative or positive, and ask if there are other ways to eliminate reproduction generations.

OVEREXPOSE WHEN SHOOTING FROM PAPER LASER OUTPUT. If you're delivering paper laser output to the printer, you might ask them to overexpose slightly when shooting film. Since they're creating negatives, white areas on your original (surrounding the halftone spots) become black on the film. Overexposing causes those areas to encroach slightly on the jaggedy-edged, toner-based spots, reducing the problem of spot gain (it can also improve the appearance of type and line-graphics). Although this tends to wash out highlights more—because small spots are more likely to disappear—that's generally less of a problem with laser output than dark areas clogging up.

MAKING PLATES

Once the plate-ready film is done, the printer uses it to expose (or "burn") plates. The process varies depending on the type of printing (offset, gravure, letterpress, or whatever), but in most cases it's a contact photographic process, so it doesn't result in much spot variation.

The plate materials available that can be imaged directly in an imagesetter (and the Heidelberg GTO-DI press and its ilk, which image plates directly on the press) promise a future without shooting negatives at all. In this setup, the plate-making step is the only mechanical one between your computer and the printed output.

INK ON PAPER

If too much or too little ink is used on the printing press, halftone spots tend to clog from spot gain, or disappear. This can also happen if the ink/water balance is off. Similarly, any one of a number of problems with the printing press can throw off tonal values in the halftone. For example, *spot doubling* (which looks a lot like severe spot gain) often happens if the press blankets are too loose. *Spot slur* occurs if the paper stretches too much or "skids" slightly across the impression cylinder. If you don't understand these terms, don't worry; unless you're a printer, you shouldn't have to, because there's nothing you as a desktop publisher can do about them (except find a new printer).

PAPER STOCK

The area in which you have to be most aware of spot variation is in the choice of paper stock. You'll always encounter noticeable spot gain on uncoated stock, because the ink is absorbed into the porous paper. An extreme case of this is on newsprint, where spots can easily gain 20 percent.

Coated stock reduces spot gain considerably, because the paper isn't as porous. Superslick paper, on the other hand, sometimes causes spot slur (which has an effect similar to spot gain) because it can slip slightly in the press. Once again, consult with your lithographer when choosing a paper stock.

We discuss paper stock in relation to screen frequency and spot gain in Chapter 5, *Setting Your Screens*.

Upper and Lower Tint Limits

There is one incredibly easy way to avoid spot variation in extreme highlights and shadows: eliminate them entirely. If no tint in your publication is too light or too dark, the problems of spot gain are somewhat reduced (but not eliminated; it's still a factor throughout the tonal range). We discuss how to manage this with scanned images using tonal compression in Chapter 20, *Tonal Correction*. In the meantime, Table 4-1 suggests tonal limits for tints and halftones on different paper stocks.

Table 4-1 Upper and lower density limits

Paper type	*Minimum tint*	*Maximum tint*
Coated	5%	95%
Uncoated	10%	90%
Newsprint	12%	88%

DTP Versus the Blob

You're probably getting the idea right about now that creating high-quality halftoned images is much harder than the salesperson at the computer store told you. Spot variation is a major issue in scanning and halftoning, though it's one rarely focused on. Fortunately, it's also one that's surmountable.

Although proper attention to paper stock, calibration, and the other elements we've discussed in this chapter will put you well on your way to creating excellent-looking halftones, one factor is so important in the discussion of spot variation that we've given it its own chapter: screen frequency. That's where we're going next.

5 Setting Your Screens

Frequency and Spot Gain

As we've seen in the last two chapters, screen frequency is one of the most important factors you need to consider when working with halftones. We looked at the tradeoffs between screen frequency and number of gray levels back in Chapter 3, *Frequency Versus Gray Levels.* In Chapter 4, *Reproducing Halftones,* we discussed the factors that cause spot variation. Now we want to look at how screen frequency affects spot variation.

The Problem with High Frequencies

The key issue is that as the screen frequency of a halftone increases, it becomes more susceptible to spot variation and muddiness. There are two reasons for this frustrating little fact. First, because spot gain occurs at the edges of spots, when there are more spots per square inch, there's more spot gain happening (more spots = more edges = more potential spot gain). Second, at higher frequencies the spots are so close together that any spot gain can fill in what tiny space there is between each spot.

But how do you know what screen frequency you can get away with? There's no hard-and-fast rule to this; rather, you need to take several factors into consideration.

Factors to Consider

Typically, unless you're trying to achieve some sort of special effect, you'll want to use the highest screen frequency you can. However, maximum screen frequency is dictated by many different factors. Generally, the rule is to ask your printer what the proper screen frequency should be; but of course that doesn't always work. We're often reminded of the time we brought some artwork to a quick-print shop where the kid behind the counter thought screens were used to keep flies out.

The primary factors you need to consider when setting your halftone screen frequency are output resolution, output method, and paper stock.

OUTPUT RESOLUTION

In Chapter 3, *Frequency Versus Gray Levels*, we explained the relationship between gray levels, screen frequency, and output resolution. If you adhere to The Rule of Sixteen—screen frequency should not exceed output resolution divided by 16—you'll always get your full complement of 256 gray levels. However, if you don't need all those gray levels, you can go to a higher screen frequency.

If you can't switch to a higher resolution device, you might output your scanned images and graduated fills—which require all 256 gray levels—at a lower screen frequency than your flat tints. That way you get a fine screen for type, lines, and boxes, and 256 gray levels for photographs and graduated fills. See Chapter 13, *Controlling Halftone Screens* and the application-specific chapters (27, 28, and 29) to find out how.

OUTPUT METHOD

Output method refers to the sort of printer you're outputting your artwork on. If you're producing final artwork on any type of toner-based laser printer, don't expect to exceed 110 lpi, even if the printer prints at high resolutions. Even better, use 65 or 85 lpi. The ragged-edged spots that result from toner-based devices are more prone to dot gain. On the other hand, the maximum screen frequency on an imagesetter is limited only by its output resolution.

Output Medium. Whether you use paper or film for your output makes a big difference in what screen frequency you use. If you're outputting to paper, you're limited to about a 110-line screen, because the printer's cameraperson can't hold a screen much finer than that. If you want a higher screen frequency, output to film.

This is also a good tip for laser printer output, especially with some of the new high-res lasers and esoteric halftoning techniques. Print on acetate instead of paper, and you can get away with a higher screen frequency.

REPRODUCTION METHOD

The third factor in choosing a halftone screen frequency is the reproduction method that you're using. Different printing methods dictate different screen frequency settings. Here's a very quick overview of some methods.

Offset printing. The limits on screen frequency particular to offset printing are dictated by your printer's press and the skill of the presspeople, and the paper stock you're using. It can range from 75 lpi up to as much as 300. Ask your printing rep for the proper frequency.

Gravure. This printing method can hold a fine screen very accurately. If you're using gravure, you're probably working with a top-notch printer, and don't need our advice to choose screen frequency. Ask your rep.

Quick printing with paper plates. You'll almost always want to use 120-lpi screens or lower.

PHOTOCOPYING. Screen frequencies up to 100 lpi are possible, and 85 lpi is pretty safe, though we've seen scans clog up at even this low frequency. Try 75 lpi with scanned images.

PAPER STOCK

Perhaps the biggest consideration when choosing halftoning screen frequency is the paper stock you're using for printing. There are hundreds of different types of paper, each with it's own amount of absorbency. However, the basic information to remember is that coated papers boast the least spot variation; uncoated papers suffer more because they're more absorbent; newsprint sucks up ink like the Sahara sucks rainwater (printing newspapers has been compared to putting kerosene on toilet paper at 200 miles per hour). Table 5-1 suggests screen frequencies for various stocks and printing methods. For in-depth information on paper, consult International Paper Company's *Pocket Pal.*

TABLE 5-1 Choosing screen frequency

Final reproduction method	*Paper stock*	*Screen frequency*
Photocopier	Uncoated	50–60
	Laser paper*	50–90
	Coated	75–110
Quick printer w/paper plates	Uncoated	50–75
	Matte bond	50–90
	Coated	75–110
Offset printing	Coated	120–150+
	Uncoated	85–133
	Newsprint	60–85
High-quality offset/gravure	Coated	150–300

**Laser paper has a smooth, matte finish, the better to hold toner.*

MAXIMIZING YOUR FREQUENCIES

The name of the game is getting the best-looking images on your pages, right? But as clearly as the game has its rules and regulations, it also has an individual personality. We can't truly sit here and tell you that you can do this or you can't do that. Last time we did that we said there was no way to produce and print a good-looking 400-lpi image from the desktop. We were wrong. When enough thought and care go into planning and executing your work, it's amazing what you can do.

6 The Glorious Spot

Spot Shapes for Quality and Special Effects

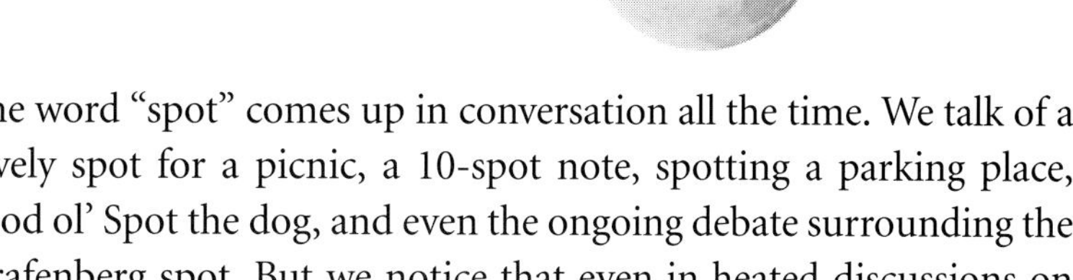

The word "spot" comes up in conversation all the time. We talk of a lovely spot for a picnic, a 10-spot note, spotting a parking place, good ol' Spot the dog, and even the ongoing debate surrounding the Grafenberg spot. But we notice that even in heated discussions on halftoning, people rarely talk about the details of the halftone spot.

We understand. In the majority of printing situations, spot shape is admittedly a minor quality factor. Halftones generally go through so many processes on the way to print—camera work, plate burning, offset blankets, paper—that carefully created elliptical spots (or whatever) might just as well be simple round blobs.

There are situations, however, where choosing a spot shape can make a real quality difference. If you're producing cosmetics catalogs, or need to solve tonal shift problems printing on newsprint at coarse screen frequencies (to use two examples), controlling the halftone spot can definitely improve the quality of your job.

On the Spot

As we mentioned back in Chapter 2, *Frequency, Angle, and Spot Shape,* you can create halftone spots in a number of different shapes. Why would you want to alter the halftone spot shape? First, you might want to create special-effect screens for design reasons. We discuss some of these effects at the end of the chapter. It's more likely, however, that you will choose spot shapes to minimize spot variation in the reproduction process, and avoid a phenomenon called optical jump.

OPTICAL JUMP

Remember those chocolate-chip cookies back in Chapter 4, *Reproducing Halftones,* and how they all merged together in the oven to form an inchoate blob? There's a point in the tint spectrum from white to black where halftone spots (and chocolate-chip cookies) start to blend together and no longer look like individual entities.

With halftones, that merging can result in an apparent jump in gray level. Figure 6-1, for instance, shows a smooth blend from black to white. At about the 75-percent point, however, where the halftone spots meet and begin to merge, there's an apparent jump in the smoothness of the blend.

FIGURE 6-1
Optical jump

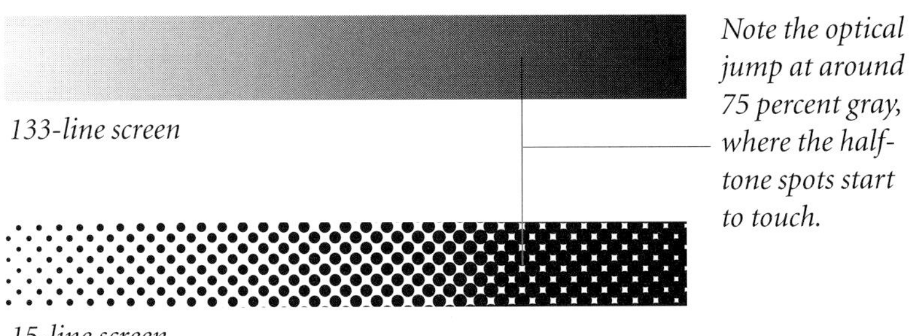

You can reduce optical jump by choosing your spot shape—especially important when you're working with graduated fills (or blends, or fountains, or whatever you want to call them) and with photographs that include smooth transitions in the midtones (as in human faces).

SPOT VARIATION

You might also want to consider spot shape because different spots fare differently in the tortuous path through the reproduction process. As we've noted before, spots tend to vary in size as they move through the process. Most spots get bigger (spot gain), though some very small spots (five or 10 percent and below) tend to disappear entirely. The shape of the spot can affect the variation because some spots retain their shapes well in highlight areas, some in the midtones, and some in shadow areas.

Spot Shapes

We'll discuss how to alter spot shapes in both the PostScript and the application-specific chapters (Parts 3 and 4). For now, let's take a look at various spot shapes and their relative merits.

ROUND SPOTS

The most simple halftone spot is round (see Figure 6-2). The spot starts small, and then gets progressively bigger until it fills the entire area of the halftone cell. The round spot fares well in light tints and highlight areas; five- and 10-percent tints with round spots are less likely to wash out and vanish. The round spot is especially prone to

Figure 6-2
Round spot

110 lpi

optical jump, however, and to serious dot gain in dark tints and shadow areas.

The first problem with the round spot is that all four sides of a round spot touch its neighboring spots all at once—at a given tint level. So at about 75 percent, where all four sides meet, there's a marked optical jump.

The second problem occurs in the dark tint areas (over 80 percent or so). The round spot that held the ink in place so well in highlight areas is now touching other spots. The technical term for the white shape between the spots (the area not covered by ink) is "that strange square pointy thing that looks like a diamond in a deck of cards." The fine, pointy areas in this shape have a tendency to fill in with ink, obliterating the subtle differences in dark tones and clogging up everything to black.

Other than the original PostScript version 23 Apple LaserWriters, there are almost no printers that use this simple round spot. However, if you don't use a PostScript output device and are relying on your scanning, image-processing, or page-layout software to do the halftoning (see Chapter 7, *Who Does the Halftone?*), there's a reasonable chance that you'll get these simple spots. You can check by printing a graduated tint from black to white with a coarse screen frequency, and examining it carefully.

THE NEW POSTSCRIPT SPOT

Starting with PostScript version 38, PostScript devices started using a variation on the round spot as their default. It actually changes shape as it gets bigger, so we call it a transforming spot (we discuss another transforming spot—the transforming elliptical spot—later in the chapter). In *PostScript Screening: Adobe Accurate Screens*, author Peter Fink calls this spot a Euclidean spot.

The New PostScript spot actually changes shape as it gets bigger. It starts out round, then at around 50-percent gray it changes shape, first becoming square, and then inverting so that instead of a big black round spot, it appears as a small white spot on a black background (see Figure 6-3). This is similar to the effect you'd get if you were creating halftones traditionally (photographically).

FIGURE 6-3 The New PostScript spot

110 lpi

20 lpi

This spot is less prone to spot gain than a normal round spot, because the round white spot in dark areas doesn't fill in with ink as easily as the strange square pointy thing. However, in many cases, it makes the optical jump problem even worse because there's a definite visual effect made by the checkerboard at around the 50-percent mark. This effect is especially noticeable in small graduated blends because the gray levels are densely packed in a small space.

OVAL SPOTS

In another attempt to solve the problems of round spots, some programs produce an oval spot (see Figure 6-4). This shape reduces the optical jump somewhat, because there are two meeting points—first where the ends meet and next where the sides meet. So instead of a single, large optical jump at 70 percent, you get two lesser optical jumps, at about 50 and 80 percent (the percentages where the ends and sides touch vary depending on how elongated the oval is).

While the oval spot has some advantages (like the round spot, it holds up well in highlight areas), it's really just a poor imitation of

FIGURE 6-4
Oval spot

the true elliptical spot, which we discuss next. It's the shape you get when you choose the Elliptical Spot option in QuarkXPress (version 3.2 and before, at least). Other programs may also use an oval spot when what you really want is elliptical. Again, you can tell what you're getting by printing out a sample at a low screen frequency (coarse enough to see the spots easily).

ELLIPTICAL SPOT

Lithographers and screen printers have used an elliptical or "chain" spot (Figure 6-5) for years because it effectively battles both optical jump and spot variation. First of all, note that the elliptical spot is not really an ellipse. It's more of a rounded-corner diamond—slightly squashed, or elongated, so it's a bit bigger in one direction than in the other. The shape address the problem of optical jump very nicely because—like the oval dot—there are two meeting points: at the ends, and on the sides. So you end up with two smaller optical shifts—at about 50 and 80 percent—instead of one big one

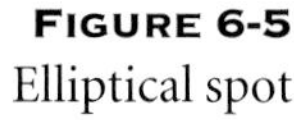

FIGURE 6-5
Elliptical spot

at 70 percent. Also, since the pointy ends of the diamonds merge slowly, the optical jump is less marked.

The elliptical spot holds up pretty well in dark areas, because the area that remains white is shaped like a diamond, as well. That shape doesn't fill in as easily as the strange square pointy that results with round and oval dots. It still doesn't hold up quite as well as a round or oval white area, however.

Note that this elliptical spot is not very elongated—it's actually close to a rounded-corner square—because if it was very elongated, you'd end up with something that looked like a line screen (or a chain-link screen, or what some lithographers call a corduroy screen) in tints between about 35 and 65 percent (read: really ugly).

TRANSFORMING ELLIPTICAL SPOT

The best halftone spots we've seen are the transforming elliptical spots that Photoshop creates when you select the Elliptical Spot option (see Figure 6-6). The transforming elliptical spot starts out

Figure 6-6
Transforming elliptical spot

110 lpi

20 lpi

as an oval, changes into an elliptical spot at around 45 percent, then inverts, appearing as an oval white spot on a black background.

Transforming elliptical spots have all the advantages of elliptical spots in that they avoid optical jumps in the midtones, but they have the advantages of oval spots in highlight and shadow areas. Light areas don't wash out, and dark areas don't clog up. The only tiny improvement we'd hope for is that there be one more transformation—from oval to round at about 15 and 85 percent, since round spots are even better than oval for holding the ink in light areas, and holding out the ink in dark areas. Well, maybe someday.

Square Spot

Square spots (see Figure 6-7) are often used in high-quality color catalog work, because they give the impression of sharpness—especially in the midtones that are so important to cosmetic photography and the like—without having to use a whole lot of sharpening on an image (see Chapter 21, *A Sharper Image* for more on sharpening and

FIGURE 6-7
Square spot

its problems). They give this impression partially because the corners of the dots suggest sharpness to our visual system, but also because they result in an optical jump in the midtones. That optical jump gives the midtones contrast—notably in human faces—and makes them look "sharp."

Square spots have also been used traditionally in photoengraving for letterpress printing. Many programs offer the option of using a square spot. Though it's mainly useful for the somewhat specialized worlds of catalog and letterpress printing, it's also useful for creating special-effect screens, which is what we cover next.

Special-Effect Spots

Up to this point, this chapter has concentrated on spots for run-of-the-mill halftoning situations—for reproducing photographs and tints using relatively high screen frequencies (50 lpi and up). However, when you get into low screen frequencies where you can actually see the spot shape, it's another world. In this world, spots can be fun, fun, fun. You can use them to produce trendy and hip designs, and to give photographs an eye-catching, stylized look.

You can use any of the spot shapes discussed above at low screen frequencies, of course, but there are some spot shapes that are really useful only in coarse screens. Let's look at a few of these popular special-effect spot shapes.

LINE SCREENS. A line screen, in which the halftone appears to be made of straight lines, is a commonly used special halftone effect. There's really nothing special about it: it's also just a pattern of spots. It's just that each spot is shaped like a little line. The little line-spots blend with each other to form long lines through the image (see Figure 6-8).

OTHER SPECIAL-EFFECT SCREENS. There are a number of other special-effect spots you can create in PostScript—triangles, donuts, diamonds, pinwheels, and so on. Figure 6-9 shows a few of these. We explain how to create them in Chapter 26, *Spot Functions.* Again, remember that there's no reason to use these at high screen frequencies. In fact, there's rarely any reason to use them at all. But it's nice to know they exist, just in case.

Also, note that some PostScript imagesetters, especially those that are called "PostScript compatible" (that is, they aren't licensed from Adobe, but imitate Adobe's PostScript) can choke on these patterns or give you bizarre results you didn't expect.

SPOTS

Altering a halftone spot shape won't help much with getting rid of a moiré problem (see Chapter 11, *Rosettes and Moirés*), or improving the appearance of a lousy scan. In fact, it probably isn't even something that most people will see. Like good typography, the spots should rarely stand out and announce their presence. But if you're interested in high-quality work or in creating special effects in your halftone images, spot shape can help.

FIGURE 6-8
Line screen

110 lpi

20 lpi

FIGURE 6-9
Special-effect spots at 20 lpi

Triangles

Propellers

FIGURE 6-9 Special-effect spots at 20 lpi continued

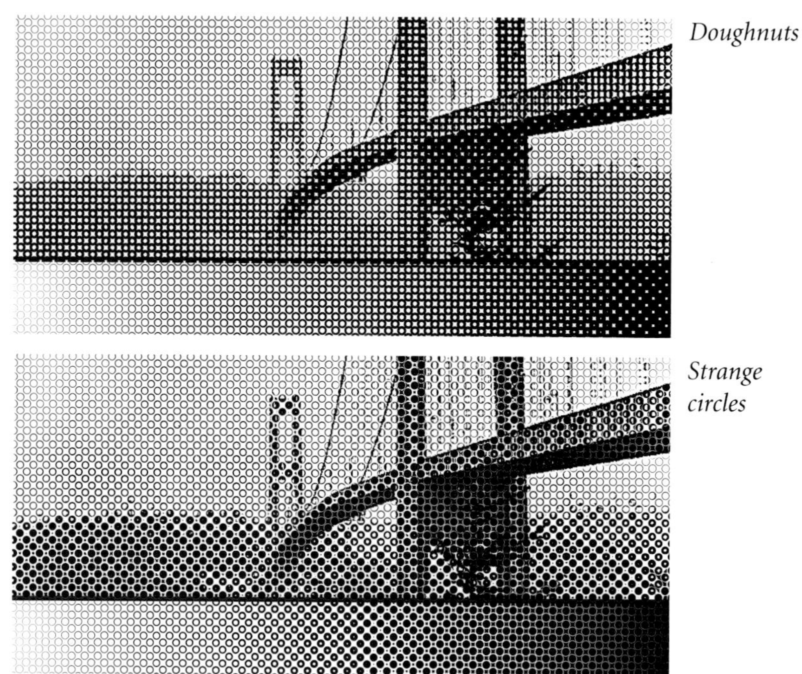

7 Who Does the Halftone?

Printers, Scanners, and Software

Steve doesn't spend much time pounding nails these days, but that wasn't always the case. One of his lesser claims to fame is that he spent one summer building a Ponderosa Steak House.

One thing he learned that summer was that the order in which you do things makes a big difference. In some situations it's better to build a whole wall lying flat, then tip it up into place. Other times, you need to build the wall vertically, assembling it where it stands. The order in which you build has a big effect on what you can build, how easy it is to build, and when you can build other things. And it's all dependent on the tools and materials you're working with.

The same is true with halftoning. In some situations you may want or need to do all the halftoning as you scan an image. In other situations, you might rely on your software to do the halftoning. Most often, you can rely on a PostScript printer to do the halftoning for you, when you print the job (see Figure 7-1).

Halftoning in PostScript

You almost never have to think about when the halftoning happens, or who does it, because PostScript manages it all for you. You simply build publications, specify tints, and place scanned images, then

FIGURE 7-1
Comparisons of halftoning at scan time, in software, and at output time.

Halftoned scan, non-PostScript printer. *When halftoning happens at scan time, the scanner delivers a flat bitmap to disk. The software can't change it at all, and passes the bitmap information directly to the printer. The problem: the halftone spots are tailored for a given output resolution and size, so you get plaid patterning at non-integral scaling percentages.*

Grayscale scan, non-PostScript printer. *The software creates the halftoned bitmap from the scanner's grayscale image. You get a result much like the one above, but you can scale it and print to a higher-resolution output device, because halftoning doesn't happen until print time.Tonal correction is also possible because of the grayscale data.*

Grayscale scan, PostScript printer. *The software passes the grayscale data through to the printer, without halftoning. The PostScript interpreter handles the screening.*

print. Your software tells the PostScript output device what gray levels it wants (along with what screen settings—frequency, angle, and spot shape), and the PostScript interpreter turns those grays into halftones using the specified screen settings.

Halftoning in Software

There are times, however, when it's advantageous or necessary for your software to do the halftoning before you get to the printer. There are two primary reasons you might call on your software to do the halftoning.

- You're not printing to a PostScript printer.
- You want to use a halftoned image for some effect, and you can't achieve the effect with PostScript's halftoning machinery.

Halftoning on Non-PostScript Devices

Non-PostScript printers, such as Hewlett-Packard LaserJets, PaintJets, and the like, don't know *bubkes* about halftoning. They certainly don't know how to build those beautiful little halftone spots. In most cases, all they understand is "turn a dot on, turn a dot off." So in order to create halftones on these devices, your software has to tell them explicitly which dots to turn on.

If you're printing from PageMaker to a run-of-the-mill LaserJet, for instance, PageMaker can't ask the printer to "make that box 90-percent gray." Instead, PageMaker has to create a big, 300-dpi, black-and-white bitmap (halftoning it in the process), and send that halftoned image to the printer. The same is true with a simple tinted box. It has to convert the box to a big 300-dpi halftoned bitmap (*rasterizing* it, or converting it to a raster—a.k.a. *bitmapped*—image) and send the whole bitmap to the printer. We call this *bit-blasting.*

Often it's not actually your application program that handles the halftoning in this situation. In many cases, the printer driver does the halftoning at the request of the application program. So you tell CorelDraw that you want a 90-percent tint. It sends that request to the Windows DeskJet printer driver; the driver creates the halftoned bitmap and sends it to the printer.

Whether it's the application or the driver that does the halftoning, you're still faced with sending a big flat bitmap to the printer. If you're printing from a PC parallel port, the transfer time isn't too much of a problem. If you're printing through a serial port, however, or over LocalTalk, the transmission time can constitute a major speed bottleneck.

The other limitation of halftoning in software is control over screen settings. Few applications (or drivers) provide the control over frequency, angle, and spot function that PostScript does.

HALFTONES AS PATTERNS

Another scenario: You want to create a halftone-like pattern for the background of some groovy ad. So you fire up Photoshop, PhotoStyler, or the like, create a gray box, then convert to black-and-white ("bitmap") mode. One of the options for the conversion is "Halftone," which lets you choose the frequency, angle, and spot shape of the resulting image. The program converts the image to a black-and-white halftoned bitmap at the resolution you specify (see Figure 7-2).

Photoshop even has a Color Halftone filter, which creates images that look like overlaid process color screens. You'll see dozens of examples of this halftoning-in-software technique throughout the book, since we used it extensively to create the illustrations.

HALFTONING AT SCAN TIME

Instead of halftoning in PostScript or in the application, you may encounter situations in which halftoning at scan time is necessary or desirable. In this situation, the scanner (or scanner software) does the halftoning, and delivers a black-and-white, prehalftoned image to disk.

Almost every scanner/software combination offers the option of scanning in black-and-white halftone—also called "dithered"—mode. Most offer a choice between halftone patterns.

FIGURE 7-2
Halftoning with Photoshop

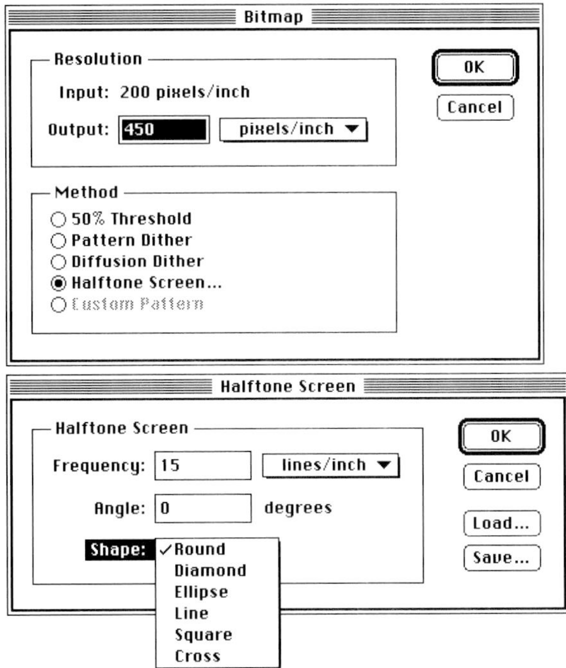

There are two situations in which it makes sense to halftone at scan time.

- Your scanner isn't capable of producing grayscale images (often true with hand scanners and older sheetfeds and flatbeds).
- Your final output is to a 300-dpi laser printer, and you can save disk space by creating a prehalftoned 300-spi bitmap rather than storing grayscale information.

NON-GRAYSCALE SCANNERS

If your scanner is limited to black and white, you're stuck with halftoning at scan time, creating a 300-dpi black-and-white halftone. This is fine as long as your final output is to a 300-dpi laser printer; you're getting the same resolution halftone; the only difference is when the halftoning happens. There is one important consideration, however, when working with images halftoned at scan time: integral scaling and patterning (see "Black-and-White Bitmaps with Repeating Patterns" in Chapter 10, *When Grids Collide*).

Deducing gray levels. Note that scanning software for several non-grayscale scanners (notably handscanners) and some image-manipulation programs can *deduce* grayscale information from the prehalftoned bitmap. These programs look at the halftone pattern, and based on the size of spots in different areas, decide that those areas should be a certain gray level.

The problem with this technique is that there simply isn't enough information for good deduction. At best, you can expect resolution in the grayscale file that equals the screen frequency of the halftoned screen in the source scan (generally about 60 lpi). Plus, the halftone pattern in the scan often comes through into the grayscale version, corrupting the image.

HALFTONING FOR LASER OUTPUT

If you're short on disk space and are printing final output on a laser printer, halftoning at scan time can reduce file sizes by a factor of eight (you're creating one-bit halftoned scans, rather than eight-bit grayscale scans). As long as you bear integral scaling in mind, you can achieve the same quality as with grayscale scans (they're 300-dpi halftones either way), but with much smaller files.

For example, a four-by-five-inch image is 1,757K (kilobytes) in 8-bit (256 grays) grayscale mode at 300 spi, but only 220K as a pre-halftoned, one-bit, 300-spi image. But if you print this image on a high-resolution imagesetter, the halftone quality is no better than it would be on the low-resolution laser printer. To achieve good results for imagesetter output using scan-time halftoning, you'd have to create at least a 1,000-spi bitmapped image, resulting in a 2,441K file.

Streamlining the Process

It's true, you might not need to know the detailed mechanics of who does the halftoning, when, and how. But the time will come (probably tomorrow) that you will. By understanding when and where halftoning happens, you can really control the process, and get the kind of results you want without swallowing your hard disks, clogging up your laser printers, and short-circuiting your imagesetter.

8 Beyond the Spot

Halftoning Alternatives

The way we're talking, you'd think that there's only one kind of halftone in the world: the kind with a frequency, angle, and spot shape, where spots are big in dark areas and small in light areas. Well, in this little chapter, we're saying that's just not so. Remember that our definition of a halftone is a black-and-white simulation of gray. But who says the grays have to be simulated with spots?

It was not that long ago that a new technology took amusement parks by storm: you could go into a booth and have your face printed on a T-shirt, calendar, or poster. Did they use ink-jets or dye-sub printers? Did they use lasers? No! Your face came out printed with teletype characters: the letters of the alphabet, numbers, and symbols. The closer the characters were, the darker the image; the farther apart the characters, the lighter the image. The halftone, although a little rough, certainly worked.

In this chapter we'll look at several less conventional methods of halftoning, including regular- and random-pattern halftones. Many of these sorts of halftones are not appropriate for high-quality graphic arts work; others are *only* appropriate for the most demanding graphic arts work.

Regular-Pattern Halftones

Normal halftones (the ones we've been talking about so far in this book) are tiled patterns. That is, they are made of a single pattern that gets repeated over and over again in a big grid, like tiles on a floor. We call these *regular-pattern halftones* because they never deviate from the regular grid.

However, there are regular-pattern halftones that aren't made up of variable-sized spots. Figure 8-1 shows an example of a halftone pattern that is tiled, but is built by turning dots on and off in a non-intuitive way. Photoshop can create custom regular-pattern halftones in any shape or size.

Regular pattern halftones are created in a fashion similar to the halftone spots we've been talking about. A pattern is always in the shape of some sort of rectangle that gets repeated over and over throughout an image. To generate dark and light areas, the dots that

FIGURE 8-1
Regular-pattern halftones

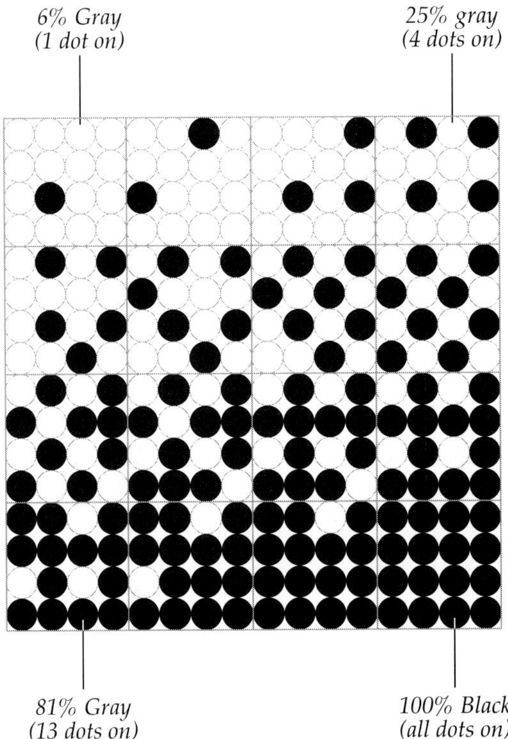

make up the pattern are turned on and off as necessary. In a dark area almost all the dots in the pattern rectangle are turned on; in a light area, almost none of the dots are turned on.

The main difference between these halftones and normal halftone spots is the order in which the dots are turned on. In a halftone spot, the pattern grows outward; in other sorts of regular-pattern halftones, the dots are turned on somewhat randomly, creating the pattern slowly.

Frequency-Modulated Screening

There is another way of creating halftones, however. Instead of using a shape that repeats on a regular grid, with a *random-pattern halftone*, the spots are all the same size. However there are more of them in dark areas, and fewer in light areas (see Figure 8-2). It's also called *frequency-modulated (FM)* screening, because the size of the spots doesn't change, just their frequency—the number of spots—in a given area. Frequency-modulated halftones look a lot like traditional photographic mezzotints (by the way, "mezzo" means "half," so "mezzotint" really just means "halftone").

The teletype halftone described earlier is an example of a random-pattern halftone, because there were lots of characters in some places and few characters in other places. In a frequency-modulated halftone, it's the *number of dots in an area,* rather than the spot size, that determines the gray level.

FIGURE 8-2
Random-pattern halftones

There are a couple of advantages to frequency-modulated screening. It avoids both pattern and content moirés in color separations; it can allow you to use lower-resolution scans; and it can be a good alternative for lower-resolution laser output (see Figure 8-3).

CREATING FREQUENCY-MODULATED HALFTONES

There are basically two ways to create frequency-modulated halftones.

DIFFUSION DITHER. Use the *diffusion dither* option in your scanning, image-manipulation, or page-layout program. This feature goes by various names. In Photoshop and PhotoStyler it's called Diffusion Dither. In QuarkXPress it's called Ordered Dither. In Digital Darkroom, it's Advanced Halftoning. This method is appropriate for lower-resolution laser output (like your desktop printer), especially if you're not going to reproduce it.

FIGURE 8-3 Traditional versus random pattern screening on laser output

STOCHASTIC SCREENING. Some of the imagesetter manufacturers' have developed imagesetters that can make "frequency modulated," or "stochastic," screens. ("Stochastic" means "random," but "Random Screening" doesn't look very good in imagesetter marketing brochures.) This method is only useful for high-end color separations.

SIZE IS (STILL) WHAT COUNTS

Even though random-pattern halftones rely on the number of spots in an area rather than spot size to depict different gray levels, spot size is still important with these techniques. After all, you still have to print them.

The problem with the stochastic screening techniques available at the time of this writing (Agfa's CrystalRaster and Linotype-Hell's Diamond Screening) is that they use very small spot sizes—around 15 microns (or about 60-millionths of an inch). The problem is that 15 microns is about the size of a one-percent dot at 150 lpi. And as we learned in Chapter 4, *Reproducing Halftones*, few printing presses can hold that fine a spot, resulting in clogged-up shadows and washed-out highlights. Therefore, these methods should be reserved for very high-quality (read: expensive) printing situations.

Neither of the stochastic screening methods offers any control over spot size, though we expect later implementations to do so. With larger spots, frequency-modulated screening might be a great replacement for traditional screens in a variety of printing situations. Nobody has really had a chance to test printing for different stochastic screening methods. Call Agfa, Scitex, or Linotype-Hell and ask them for information or print samples.

ALTERNATIVE ALTERNATIVES

It's not clear whether alternatives to traditional screening such as random-pattern halftones will really be important for most desktop publishers. However, we do think that stochastic screening could well be a major step in color work because it avoids the bane of a color separator's existence: moiré patterns. We cover these troubles in Chapter 10, *When Grids Collide.*

9 Band Aid

Fountains, Blends, and Vignettes

"Why can't I get a smooth graduated blend? Why are there all these bands in my fountain? How can I get rid of this shade-stepping?" These questions get asked everyday, everywhere that people do desktop publishing. The problem is easy to see (see Figure 9-1), especially in graduated fills that extend over a large area. And it's usually almost as easy to fix.

Before we go any farther, for the sake of everyone's sanity, let's agree that the terms fountain, blend, degradé, vignette, and gradient (or graduated) fill all refer to the same thing. Some people fight over what they consider to be the best term, but finally it seems to come down to religious preference. Whatever you want to call it, this kind of fill is very popular among desktop publishers, especially among those doing illustrations.

We'll start out this chapter by looking at how banding occurs, then talk about how blends are made and how you can create them so as to avoid banding.

Figure 9-1
Banding in gradient fills

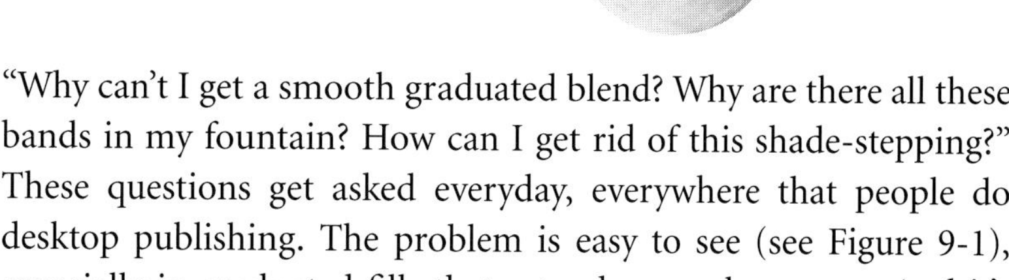

Running from 30 percent to 50 percent gray over 34 picas, this blend displays noticeable banding.

SHADE-STEPPING

Banding typically appears in blends for one of two reasons: uneven motion of the imagesetter transport mechanism, or the inherent limit in gray levels of digital halftoning. The artifacts left by uneven transport are more accurately called striations (or "stripes"), and they're not nearly as much of a problem today, with modern imagesetters, as they were in the good old days (back when nothing really worked properly).

TRANSPORT MECHANISMS

PostScript imagesetters consist of two parts: the raster image processor (RIP) and the recorder. The RIP takes image information from the computer, turns it into a giant bitmap, and passes that on to the recorder. The recorder images that bitmap onto film or paper. If the recorder is faster at imaging than the RIP is at processing the information and passing it along, the recorder has to image a little, then stop and wait, then image a little more, then stop and wait, and so on.

If the mechanism that transports the film or paper past the laser isn't perfectly registered, images might be slightly compressed or expanded at the places where it stopped. This slight aberration can be very noticeable with halftoned images and graduated fills.

Note that this problem only occurs with imagesetters that pass the paper or film past the laser, and even with those, most recent devices have largely eliminated the problem. Many new imagesetters are drum-based; the paper or film remains stationary on a cylinder while the laser moves. This is a significantly more precise method (but it has its limitations, too). A good test to see if your striping problem is hardware-based is to print your image out in both portrait and landscape orientations (horizontal and vertical). If the banding is perpendicular to the film in both cases, then the problem is probably in the imagesetter. If the stripes look the same in both orientations, however, then the problem is yours.

To solve a hardware-based problem, you (or your service bureau if you don't own or operate an imagesetter) need to repair the transport mechanism, upgrade to a faster RIP, or use an imagesetter that doesn't have this sort of problem.

TOO FEW GRAY LEVELS

The most common shade-stepping problem happens for much more mundane reasons. As we discussed back in Chapter 3, *Frequency Versus Gray Levels,* there are only so many levels of gray that you can achieve at a given screen frequency and output resolution. For example, if you're printing at 133 lpi on a 1,270-dpi imagesetter, you cannot achieve more than 92 levels of gray. So, when you're printing a nice, smooth blend that contains all 256 levels of gray (remember that PostScript only understands 256 different shades of gray), you're still only getting 92 of those grays. If the blend extends over a large enough area, each step is large enough that you can see it clearly. It looks like 92 bands of gray.

What can you do about this? You've got two choices: lower the screen frequency of your halftones, or print on an imagesetter with a higher resolution (or both). Either way, you raise the number of potential gray levels, and reduce banding.

Let's look at the example above again. If we change the screen frequency from 133 to 80, we can achieve 252 levels of gray at 1,270 dpi—reducing banding considerably. Or, if we print the higher screen frequency on a 2,400-dpi imagesetter, we'd theoretically be able to print 326 levels of gray (of course, in reality we'd still only get the maximum 256 levels).

TO MAKE A BLEND

How you create a blend can have a big impact on whether you get banding, and how much. There are (at least) three different ways to create blends in desktop publishing.

- Use the blend option in the Fill dialog box of an object-oriented drawing or page-layout program.
- Use the blend (or similar) command in a drawing program to create a number of intermediate objects between two objects.
- Use the blend (or similar) command in a paint or image manipulation program.

BLENDING IN FILL DIALOG BOXES

Some programs, such as Aldus FreeHand, do all the work for you; you just say "fill this box with a graduated blend from black to white" and it does it. Figure 9-2 shows FreeHand's Graduated fill option. Other programs, such as QuarkXPress, Adobe Illustrator, and CorelDraw, use similar techniques.

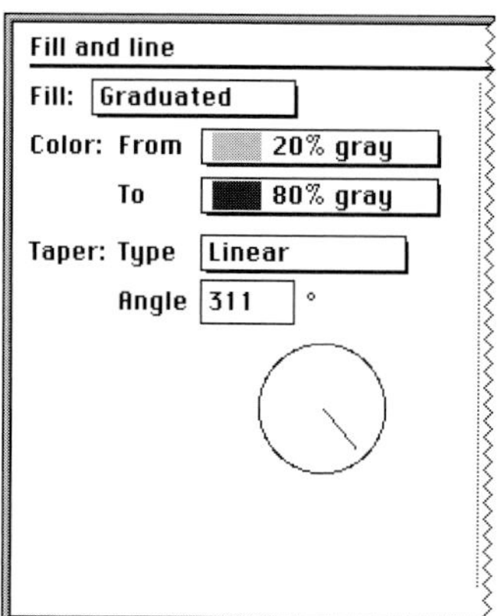

FIGURE 9-2 The Graduated fill option in FreeHand's Fill and Line dialog box

If you use a graduated fill from a program's fill dialog box, the program should calculate the proper number of steps needed to print the blend without banding (given the problems of imagesetters and digital halftoning outlined earlier in this chapter). Some programs, such as Aldus FreeHand, let you choose between a linear and logarithmic fill. Because of the way that the eye perceives light, logarithmic fills often appear to have a smoother transition from black to white. However, we still typically use linear blends.

BLENDING BY CREATING ADDITIONAL OBJECTS

Some programs also give you tools for creating gradients a different way: by blending two objects. You select the two objects containing the beginning and ending tints, tell the program how many steps you want in between, and it creates that many new objects between the two end points (see Figure 9-3).

FIGURE 9-3
Creating a blend between two objects

But how many steps should you use between the objects? The following formula makes it easy.

Number of steps = (output resolution ÷ screen frequency)^2 × percent change in color

We know this might be a daunting equation at first, but let's look at why it's really pretty easy. The first part of the equation is the same as the one we looked at back in Chapter 3, *Frequency Versus Gray Levels.* By dividing the output resolution by the screen frequency and then squaring that number, you find the number of gray levels that you can possibly print. Because PostScript can print no more than 256 levels of gray, any value above 256 should just be rounded down to 256. Next, by multiplying that value by the percent change in your blend, you find the number of gray levels that are possible in your particular blend.

For example, if you have a blend from 10-percent to 90-percent gray, and you're printing it on a 2,540-dpi imagesetter at 150 lpi, you know that your blend has a maximum of 204 gray levels possible (80 percent of 256 levels, or 256 multipled by 0.8). Because this is the maximum number of grays you could possibly achieve in this situation, there's really no need to use more steps than that in the blend (each step has it's own gray level).

If the resulting number of steps results in gaps between objects, feel free to add more steps or make the objects wider. Otherwise, your blend will look like garbage. This formula simply ensures that you have *at least* enough steps to take advantage of PostScript's 256 possible gray levels.

HOW BIG THE STEPS ARE

There are two factors to consider when figuring out whether your blend will contain shade-stepping. First, if your blend doesn't contain about 256 levels of gray, there is an increased likelihood that you'll see banding. Note that this banding is much less visible at coarser screen frequencies. However, often a bigger problem is how big each step is. Imagine 256 steps of gray from black to white over a one-inch space. Each step is going to be really tiny, because each step is only 1⁄256 of an inch.

Okay, now stretch your imaginary blend out to 20 inches. Now each step has to be much bigger to fill the gap. You can figure out how big one step is by dividing the total length of the blend by the number of gray levels possible for that blend (which we just figured out in the previous section). For example, if you know that you only have 204 levels of gray possible for a blend over 10 inches, then you know that each step will be .05 inch (10 inches ÷ 204 steps).

If the step is big enough, then it becomes obvious next to its neighbor steps. Typically, values between .01 inch and .03 inch (.25 to .75 millimeters, or one to two points) are small enough to blend together well. In the example above, the .05-inch step may be too large for a smooth blend, especially since you are not achieving a full 256 levels of gray.

COLOR BLENDS

Remember that when you separate process colors, each plate becomes in essence a black-and-white image. That means that shade-stepping can and does occur on each plate separately. Therefore, when you're figuring out how likely it is that you'll get banding, you need to look at each color plate. Typically, banding is worst in the color with the smallest percent change. For example, if you're blending a color made of 70-percent cyan and 40-percent magenta to white, then the magenta plate is more likely to band than the cyan because it's only using 40 percent of the available grays (see the equation earlier in this chapter).

On the other hand, if banding occurs on the yellow plate, chances are that it won't appear as prominently as it would on the cyan plate. Also, color blends are sometimes less likely to band because each color can mask the aberrations of the other. That is, the sum of the colors is smoother than each of the parts.

BLENDS IN BITMAPPED IMAGES

Programs that work with bitmapped images, such as Adobe Photoshop or Aldus PhotoStyler, create blends in a different way. The interface is similar: you tell the program that you want a blend from this color to that color, and it goes and does it for you (see Figure 9-4). However, because the program works with bitmapped images rather than objects (see Chapter 16, *Scanned Images*), the blend is described as a giant grid of dots, each of which can have a different gray level.

FIGURE 9-4
Using Photoshop's Blend Tool

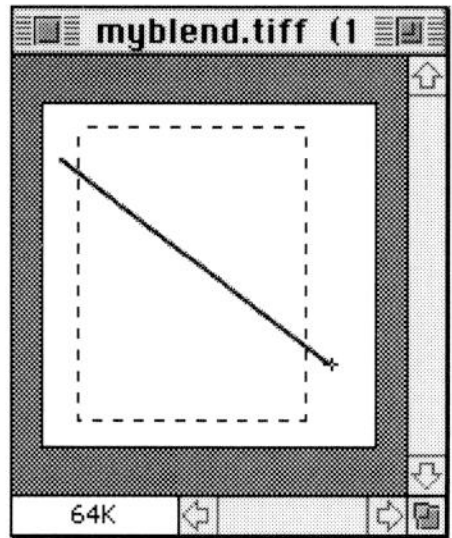

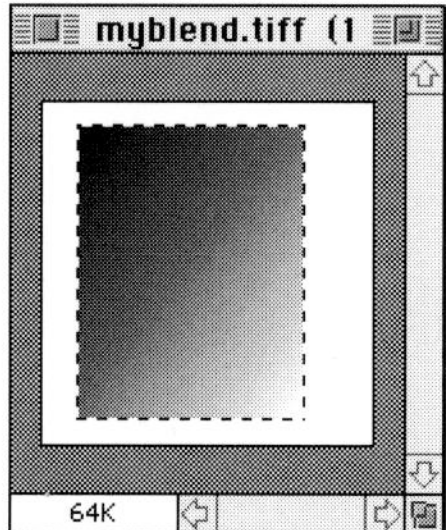

With bitmapped blends, the smoothness of the gradation is mostly dependent on the resolution and number of gray (or color) levels in the bitmap. If you're working with an 8-bit (256-gray) file or 24-bit (16-million-color) file, and your image resolution is at least equal to halftone frequency, you should be fine.

There is a real advantage to creating blends in a bitmapped program as opposed to an object-oriented program: the ability to add "noise" that disguises banding. If you include some pixels in each gray level that are of a slightly different gray level, your eye merges all those pixel together, resulting in an apparently smoother blend (see Figure 9-5).

FIGURE 9-5 Adding noise to a blend

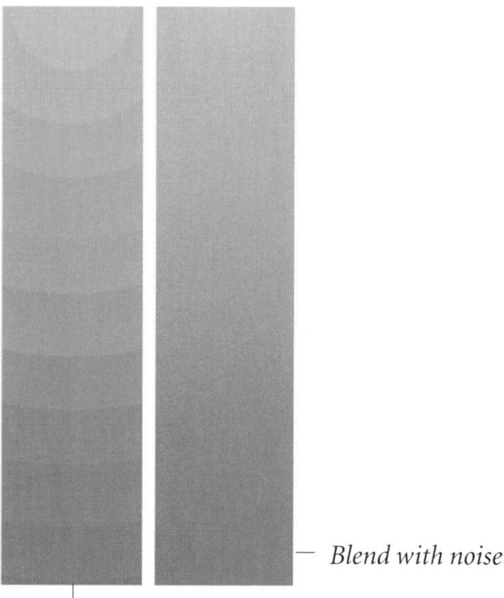

Some output devices (such as those using Varityper's ESCOR screening) can add noise to their gradient blends automatically. If you're not using one of these, you have to add it manually in Photoshop. The best way to do this is to select the gradient area and use the Add Noise filter. In typical images, where the image resolution is about 1.5 or 2 times the halftone frequency, we suggest using a noise level of about 4 or 5. If the ratio is less, you might have to reduce the amount of noise.

Banding and Reproduction

Believe it or not, banding is one area where your final printed output may look better than what you see on screen. We've seen many situations where banding is obvious and annoying in Chromalins or Matchprints, but completely invisible in the final printed output. The vagaries of printing mechanics—notably the spot variation we discuss in Chapter 4, *Reproducing Halftones*—actually tend to mask the banding, blurring the bands together so they're much less noticeable.

So, where there's a lot of spot variation—on uncoated stock and newsprint, with colored papers, and on web presses—this "band-blurring" is more pronounced. It's not a sure thing, however; sometimes you'll still get the banding effect even in these circumstances.

Getting Great Blends

When it comes right down to it, you probably shouldn't try to create blends larger than about six inches using an object-oriented illustration program like Illustrator, FreeHand, or CorelDraw—at least not in high-quality printing situations. If you expect more band blurring in the print job, you can go up to nine or 10 inches. On the other hand, smaller blends work fine with these programs. If you need blends larger than that, create them in a bitmap-editing program, adding a little noise to smooth out the banding.

10 When Grids Collide

Avoiding Patterns and Moirés

Chain link fences. Porch screens. Fishing nets. All around us are examples of big grids made of little, repeating squares. There's a fascinating optical illusion that happens with screens like these: it's called a moiré ("mwah-RAY") pattern (see Figure 10-1).

Moiré patterns are caused by our eyes' perception of straight lines and repeating patterns. Any time you overlay a bunch of straight lines or grids on top of each other, you have the opportunity for moiré patterns. One grid, all by its lonesome, never has a problem with these patterns. It's when you have two or more overlaid on each other that the wires in your perceptual system get crossed, and these little gremlins start to appear.

One way to think about moiré patterns is by thinking of a pool of water. If you drop one stone in the water, you get a radiating circle of concentric waves. If you drop *two* stones in the water, however, you get a complex pattern of high points (where two waves meet) and low points (where two troughs meet). Moiré patterns are based on the same kind of *standing-wave* or *interference* pattern.

With scanning and halftoning, we're often faced with the kind of multiple overlaid grids that tend to result in moiré patterns. What are the halftones, after all? Big grids of spots. What is scanning? It's turning a real-world picture into a grid of sample points. What is laser printing and imagesetting? It's laying down a grid of black dots. Everywhere we turn in the scanning and halftoning world, we find

FIGURE 10-1
Moiré patterns

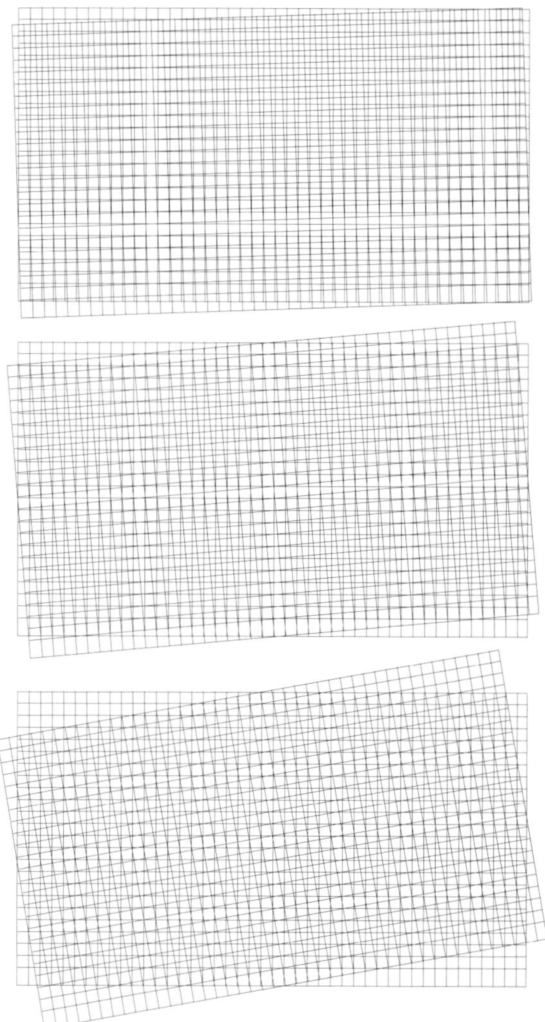

ourselves looking at grids. And, therefore, we all too often find ourselves looking at moiré patterns.

There are four main situations in which you might run into the problem of conflicting patterns.

- Printing color separations, which in most cases involves overlaying two or more halftone screens
- Scanning and printing previously halftoned images

- Printing halftones of grayscale and color images that include regular patterns
- Printing black-and-white images that include repeating patterns

Color Separations

By far the most common cause of moiré patterns is improper screening in color separations. If the angles and frequencies of the overlaid screens aren't just right, moiré patterns result (see Figure 10-2 and Color page B).

Figure 10-2
Moiré patterning in process separation

We discuss moirés in color separations in Chapter 11, *Rosettes and Moirés*, and Chapter 12, *Frequency, Angles, and Moirés.*

Halftoning Halftones

If you see a photograph in print, it's been halftoned. That's (almost) the only way to print it. You know that, but let's never forget that computers are stupid. Scanners can't tell if an image is already halftoned; they just scan it in, sample by sample, as though it were a normal photograph. There are two opportunities for interference patterns here: between the original screen and the scanner sample grid, and between the original screen (which is picked up by the scan) and the output screen.

Fortunately, there are a few things you can do to reduce the patterns in the scanned image.

FIXING THE PATTERNING

You can sometimes reduce the patterning in these cases by adjusting the image scaling and output screen frequency, using trial-and-error until you get it to work. The idea is to have an integral relationship between the original's screen frequency and the output frequency (1:1, 2:1, or whatever). If the screen angles don't match, however, that's another potential source of patterning. Unfortunately, that's a pretty darn hard thing to adjust for. A better solution (though still not perfect) is to eradicate the original halftone pattern before you print, using Photoshop, Ofoto, or some other program.

PHOTOSHOP. When our friend Jeff McCord agreed to produce the Seattle International Film Festival program, he didn't realize that the artwork he need to scan included many images that were pulled from previously printed material, even clippings on newsprint. If he learned anything from the job, it was how to scan, clean up, and print previously halftoned photographs using Photoshop.

His favorite method for grayscale images is to apply the Despeckle filter in Photoshop before anything else. Then you can adjust levels and apply unsharp masking (see Chapter 20, *Tonal Correction* and Chapter 21, *A Sharper Image*). The Despeckle filter is excellent for removing the weird traces of patterns in the scanned images. On the other hand, you have to be very careful with the Unsharp Mask filter (and other sharpening techniques) because they can re-accentuate the halftone pattern in the scan.

For scans of color halftones, Jeff likes to use Photoshop's Median filter with a specification of no more than 2 or 3 pixels. This filter has much the same effect in evening out the moiré patterning, but he finds it works better on color images.

Other people use other methods. Blurring filters followed by sharpening often work well to remove some patterning. However, you lose detail, which you can't get back by sharpening. Similarly, subsampling the image (reducing its resolution) to as low as 25 percent of original size and then supersampling it (interpolating samples) back up to the original size has much the same effect, and similar problems.

Note that people are often concerned with patterning they see on the screen with scans of previously halftoned images. However, the thing to remember is that the patterns you see at a 2:1 screen ratio may go away when you look at the image at 1:1. This is simply a conflict between the original screen and the monitor's pixel grid. The 1:1 screen ratio is the one you should pay attention to most.

Ofoto. Ofoto, from Light Source, is an incredible scanning program for the Mac and the PC that has all sorts of "intelligence" built into it about how to make good-looking scans. One of the things Ofoto knows about is moirés and how to get rid of them.

Although it's not the fastest technique on the planet, Ofoto's Auto Moiré Removal has the great benefit of doing all the work for you. In fact, it does it so well that there's even a rumor going around that Ofoto's capabilities in this area are actually built-in magic rather than true science. It works especially well with halftoned images that have screens below 60 lpi and above 110 lpi. In between, you still get some patterning.

Content Moirés

A less common problem, but one which we think you should at least be aware of, arises when you're scanning images that themselves include repeating patterns—the tight weave in a closeup of a silk tie, or the mesh screen on the front of an electric heater, for instance. These patterns can interfere with the grid of the printed halftone screen, resulting in *content moirés*. Fortunately, content moirés don't happen often, and they're often not important if the object only has a small role in the picture.

You can correct for content moirés by adjusting the image size or the output screen frequency, once again seeking to establish an integral relationship between the image pattern and the output screen frequency. Unfortunately, with content moirés it's a trial-and-error process. Stochastic screening is another way to reduce content moirés (see Chapter 8, *Beyond the Spot*).

Content moirés can also be a problem when printing grayscale or color screen shots from Windows or the Mac, which include repeating patterns in their scroll bars and other interface elements. Best bet: adjust the interface colors to get rid of patterns before you grab the screen shots.

Black-and-White Bitmaps with Repeating Patterns

Any black-and-white (or bilevel) bitmap that includes a repeating pattern can cause patterning, because the pattern conflicts with the output screen. Bilevel screen shots from Windows and the Mac are especially subject to this problem, due to the repeating patterns in their interfaces (see Figure 10-3).

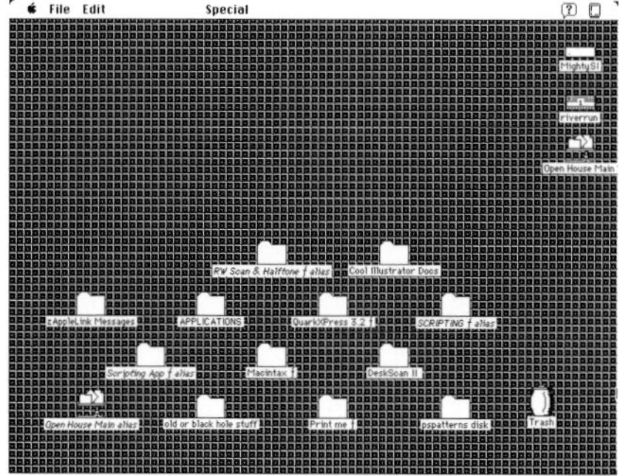

Figure 10-3 Patterning in a black-and-white screen capture

The worst offenders, however, are images that are halftoned at scan time (see Chapter 7, *Who Does the Halftone?*). These scans—which are just big grids of black-and-white dots grouped into a regular pattern of halftone cells—result in a special kind of patterning. We call this—for want of a better name—*plaid patterning* (see Figure 10-4).

Whether it's screen shots or scanned halftones, the solution to patterning in bilevel bitmaps is to scale the image so image resolution has an integral relationship to output resolution. That is, if you're printing on a 300-dpi printer, your images (at their final size, *after* they're placed on a page and scaled) should have a resolution of 100 dpi, 150 dpi, or whatever. There should be a 1:1, 2:1, or similarly integral relationship between image samples and printer dots.

This requirement imposes limitations on the scaling percentages you can use for these types of images. The best advice is to decide in advance exactly what size you need on a page, then scale and crop images as you scan them, rather than scaling in your page-layout software. However, if you've already got the scans and need to make them fit your spread, you can calculate the acceptable scaling percentages using the following formula.

Scaling Percentage = Original Image Resolution ÷ Output Resolution × Any Integer

Output resolution in this formula is for your *final output device*, not for your proofing device (if they're different, that is). You may still get plaid patterns on your proof output, but you won't when you imageset the file.

By substituting different integers in this formula, you can build yourself a little table of acceptable scaling percentages for a job. Since different programs have different levels of percentage scaling

FIGURE 10-4
Plaid patterning in a black-and-white halftone

accuracy, it's a good idea to test some images at different scaling percentages on your final output device.

If you're working with PageMaker, you can use its magic stretch feature: hold down Command (Mac) or Control (Windows) while scaling an image. This scales images automatically to acceptable percentages, avoiding plaid patterns. When you use magic stretch, images jump to acceptable sizes as you drag. Make sure to first select the printer you will be using for your *final output* as your target printer. That way PageMaker knows the final output resolution.

Love Those Patterns

We've spent this whole chapter telling you how to avoid patterns that result from overlaid grids. But there's another option: you could simply grow to love them, as David has. Of course, there's still the problem of getting your customers or clients to love them, as well.

11 Rosettes and Moirés

Beauty and the Beast

Up to now, we've been talking mostly about grayscale images. But if you're not working in color now, there's a very good chance you'll be doing it soon. Everyone is moving toward color at warp speed (even *The New York Times* and *The Wall Street Journal,* bastions of black and white). While there are books that cover color work in much more depth than we do here, we do want to take a quick detour to discuss printing process colors.

In this chapter, we talk about how halftone screens interact when you make color separations and overlay them on press. These overlaid screens are what provide the flexibility of color printing, but they can also pose real moiré problems. Moirés are mainly a problem with process color work, in which four (or even more) screens might be overlaid, but they are also a consideration with spot-color work (especially with duotones and tritones), where two or three screens might collide.

Minimal Moirés

As we said in the previous chapter, *When Grids Collide,* when you overlap two or more grids, you'll likely create a moiré pattern. The pattern is the result of our eye's and brain's tendency to see lines and patterns even when we don't want them to. However, there are some angle combinations that minimize the optical illusion.

The larger the difference in angle between two overlaid grids, the smaller the resulting pattern, and the less apparent it is (see Figure 11-1). So if one grid is at zero degrees and another is at 5 degrees, there is a very apparent pattern. If you rotate the second grid to 45 degrees, the pattern shrinks to the point that—at a sufficient viewing distance—it seems to disappear.

Note that a 90-degree grid is essentially the same as one at zero degrees, just as a 135-degree grid is the same as a 45-degree one (though with asymmetrical spot shapes, the orientation of the spots varies around the full 360-degree arc—not a major factor in moiré patterning). So the largest angle difference possible between two overlaid screens is 45 degrees. If you're creating duotones for a two-color job, the angles of the two color screens should be offset by 45 degrees (put the dominant color at 45, because it's least apparent; the secondary color—often black—should be at zero).

Continuing with this logic, the largest angle offset we could hope for between three screens is 30 degrees (90 ÷ 3). Angle offsets smaller than this produce moirés that are too large, hence too apparent.

But, hey: There are four process colors, not three. If you want the angle offsets to be at least 30 degrees, what do you do? Through years of trial and tribulation, the printing industry has standardized on a combination of four halftone angles: three at 30-degree offets, and one 15 degrees off. Cyan is at 15 degrees, black at 45, magenta at 75, and yellow at zero.

Because yellow is the lightest and least noticeable color, it can safely be set to zero degrees, even though zero degrees is a highly-noticeable angle, and it's only 15 degrees from it's nearest neighbor. Note that cyan is sometimes set at 105 degrees; however with symmetrical spots it's really the same thing as 15 degrees (and even with asymmetrical spots, it doesn't make much difference).

When you overlay the four process colors using these angles, the resulting moiré patterns are as small as they can be. If these angles are off even slightly, however, you can run into big-time patterning problems. We discuss the problem of getting accurate frequencies and angles in the next chapter, *Frequencies, Angles, and Moirés.*

FIGURE 11-1
Moiré patterns with different angle offsets

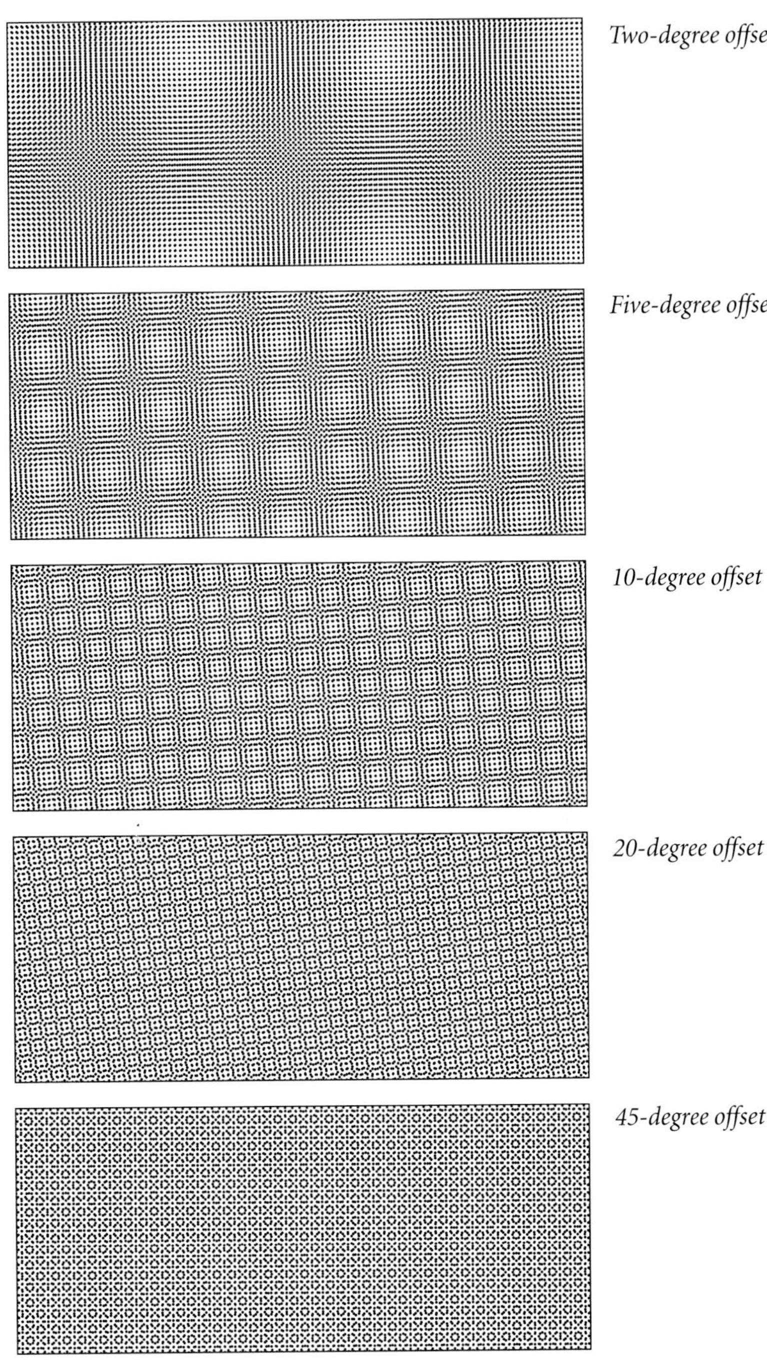

The Rosette

It's a popular misconception that when you overlay four halftone screens at the angles we just described, you don't get any patterning. You do, in fact, produce a pattern. It's just that the pattern is small and relatively innocuous, to the point that people don't call it a moiré; it's called a *rosette* (see Figure 11-2).

Figure 11-2
The rosette

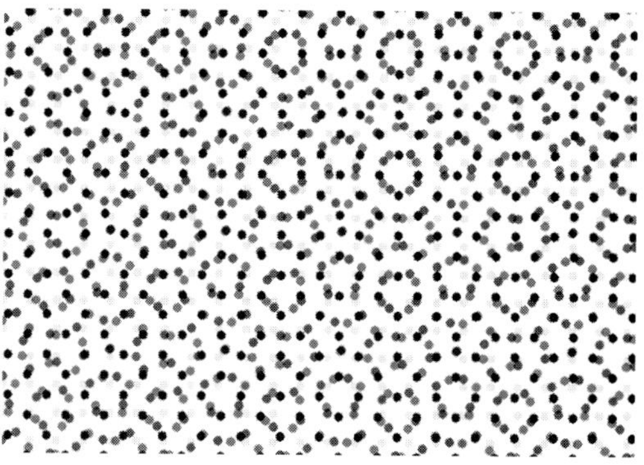

At low screen frequencies, the rosette pattern is quite evident as any pattern would be. But at higher screen frequencies, the thousands of rosettes blend together to create a smooth "surface" for the image. At a sufficient viewing distance, the patterns blend together to give an impression of photographic detail.

There are two sorts of rosettes you can make in a four-color image: open-centered and close-centered (see Color page A).

Open-centered. Open-centered rosettes don't have a spot in the center of the rosette. The pattern is sometimes called *robust* because it resists color shifts in the image even when slight misregistration occurs. On the other hand, images with open-centered rosettes display a more visible pattern, and tend to be somewhat lighter than similar images produced with closed-center rosettes at lower screen frequencies. This makes sense because so much white paper shows through.

Because of the more pronounced pattern in open-centered rosettes, they are most appropriate for use at higher screen frequencies where the patterns are small and hard to see—150 lpi and above.

Agfa's Balanced Screening produces open-centered rosettes (but it can be coerced into producing closed-center rosettes by setting the imagesetter to positive and the imaging application to negative).

Closed-centered. A closed-center rosette pattern lets much less paper show through, so you have a better chance of color brilliance, and the rosette pattern is less apparent, so it's more usable at lower screen frequencies. However, registration is more critical; minor misregistration can cause significant color shifts. Linotype-Hell's High-Quality Screening (HQS) produces closed-centered rosettes. Varityper's ESCOR screening offers the option of open- or closed-center rosettes. Note that a closed-centered rosette becomes an open-centered rosette in the highlights of an image.

Moiré Reducers

Moiré patterns can easily appear in your four-color images due to even minor flaws in either halftone angle or frequency. We talk about how to reduce those patterns in the next chapter, *Frequencies, Angles, and Moirés*, but we'd like to throw out just a few more ideas here, that don't relate directly to accurate frequencies and angles.

Screen swapping. One of the most common ways to reduce patterning is to exchange the halftone angles for the four process colors. For example, exchanging the cyan and yellow angles—so that cyan is at zero degrees and yellow is at 15—can help reduce moiré in orange colors, such as tanned Caucasian flesh tones. However, it may accentuate the pattern in purples. This process is called *screen swapping*, and it's more commonly done in traditional color houses.

Upping yellow's frequency. Some people find that increasing the screen frequency of the yellow plate can reduce moiré patterns considerably. Typically, a value of 108 percent of the other screen frequencies is suggested. For example, if the cyan, magenta, and black plates are set at 133 lpi, you might try setting the yellow plate to 143.64 (133×1.08).

FLAMENCO SCREENING. A few years back a group in Spain developed a color screening technique called Flamenco Screening in which all the screens are at the same frequency and angle, but are offset slightly. While there was great enthusiasm for the technology initially, it faded as it became apparent that flamenco-screened images were subject to large color shifts with even slight misregistration. Flamenco screening still has a niche in the newspaper market.

STOCHASTIC SCREENING. In Chapter 8, *Beyond the Spot*, we mentioned that stochastic, or frequency-modulated screening could be a viable alternative to traditional halftone screening. One of the great benefits of stochastic screening is the impossibility of moiré patterning: there are no angles or frequencies to speak of, so no patterning can occur. Stochastic screening is generally still considered to be an option only for very high-quality printing situations.

COLOR HALFTONES

Although many imagesetters now effectively eliminate most of the patterning problem through systems such as Linotype-Hell's High-Quality Screening (HQS) and Agfa's Balanced Screens, it's still important to be aware of the potential for moiré patterns. Printing four halftones on top of one another is only one of many possible pitfalls in working in color, but it's often one of the most complicated and frustrating aspects of the process.

12 Frequencies, Angles, and Moirés

You Can't Always Get What You Want

Wasn't it the Rolling Stones who told us that "you can't always get what you want, but if you try sometimes, you just might find, you get what you need"? The Rolling Stones obviously never had to work with digital halftones.

The curious paradox with digital imaging is that although it allows incredible precision—a precision well beyond what the eye can see—the precision results in significant limitations in the real world. The first limitation we looked at was back in Chapter 3, *Frequency Versus Gray Levels*: the higher your halftone screen frequency, the fewer gray levels you can achieve at a given resolution.

In this chapter, we're going to look at some other limitations: attainable angles and screen frequencies. The fact is, not only do you often not get what you want, but you sometimes can't even get what you need. For example, if you ask for a 60-lpi, 45-degree screen on most 300-dpi laser printers, you'll actually get a 53-lpi screen. If you ask for the same screen but with a 44-degree angle, you'll get the 60-line screen, but the angle snaps to 37 degrees.

We'll explain why, why it matters, and what you can do about it in the course of this chapter.

WHO CARES ABOUT ATTAINABLE VALUES?

Before we jump into looking at what halftone angles and frequencies you can and can't achieve and why, let's think about why anyone would possibly care. The main reason is color separation.

In Chapter 10, *When Grids Collide*, we explored the world of moiré patterns and how they can ruin your life. Grids over grids, we learned, can cause horrendous patterns in your artwork. And in Chapter 11, *Rosettes and Moirés*, we talked about the problems of moirés that are particular to color work. This chapter addresses color moirés, and the special considerations of angles and frequencies associated with PostScript halftone output.

When it comes to avoiding patterns, the most important attributes of a halftone are its angle and frequency. Keep those to their proper values, and moirés can often be avoided. But let them wander off to other values, and your image looks as bad as a teenager's facial complexion just before a big date.

Typically, if your angles are within .001 degrees and your frequency is within .01 lpi of the optimal values discussed in the last chapter, you'll not have patterning problems. As we'll see, this is often impossible. It depends on your equipment.

REQUESTING ANGLES AND FREQUENCIES

In earlier chapters, we've shown how you can specify a halftone screen in all sorts of ways: a given angle, a particular screen frequency, often even a special spot shape. However, due to the limitations of the digital world, you don't always get what you ask for.

Remember that the halftone spots are built inside halftone cells, which are made up of device dots. In typical digital screening, each of these cells looks exactly the same. They're like complex patterned tiles, and they have to fit together seamlessly. But there are only certain frequency/angle combinations at a given resolution where this seamless tiling is possible.

CHANGING THE ANGLE AND FREQUENCY

In order to rotate a halftone screen, you have to rotate the cell. Similarly, if you want to lower or raise the screen frequency, you have to make the halftone cells larger or smaller (see Figure 12-1). The limitations occur because the halftone cell must fall on exact printer dot boundaries (see Figure 12-2), and the cells have to tile.

Because cells have to tile, there are only so many frequency/angle combinations at a given resolution. If the frequency/angle combination that you request isn't available, PostScript uses its best guess at the nearest approximation (See Table 12-1).

INCREASING PRINTER RESOLUTION

Although 1⁄300 of an inch seems pretty small, when you're trying to build halftone screens with dots of that size, it's like building an oak dining room table with only two-by-fours. If you make the dots smaller, for instance by using a high-resolution imagesetter, you not only get better-looking halftone spots, but you increase the number of available frequency/angle combinations. When working at a finer resolution, you can rotate or adjust the size of the halftone cell, and have more possibilities of whole dots to snap to (see Figure 12-3).

If you had an infinite number of dots, you could have any combination of frequency and angle you wanted. However, neither David or Steve has had much experience with the infinite.

RATIONAL VERSUS IRRATIONAL SCREENING

All the halftone screening that we've talked about in this book so far is called *rational tangent screening*. We also want to talk about *irrational* tangent screening. Both of these terms have been thrown around in the press a lot, so we'd better clarify them.

These names derive from the sort of "snapping to the nearest dot" that we've been talking about here. When you can describe an angle by saying "move over eight dots and up three dots," you're de-

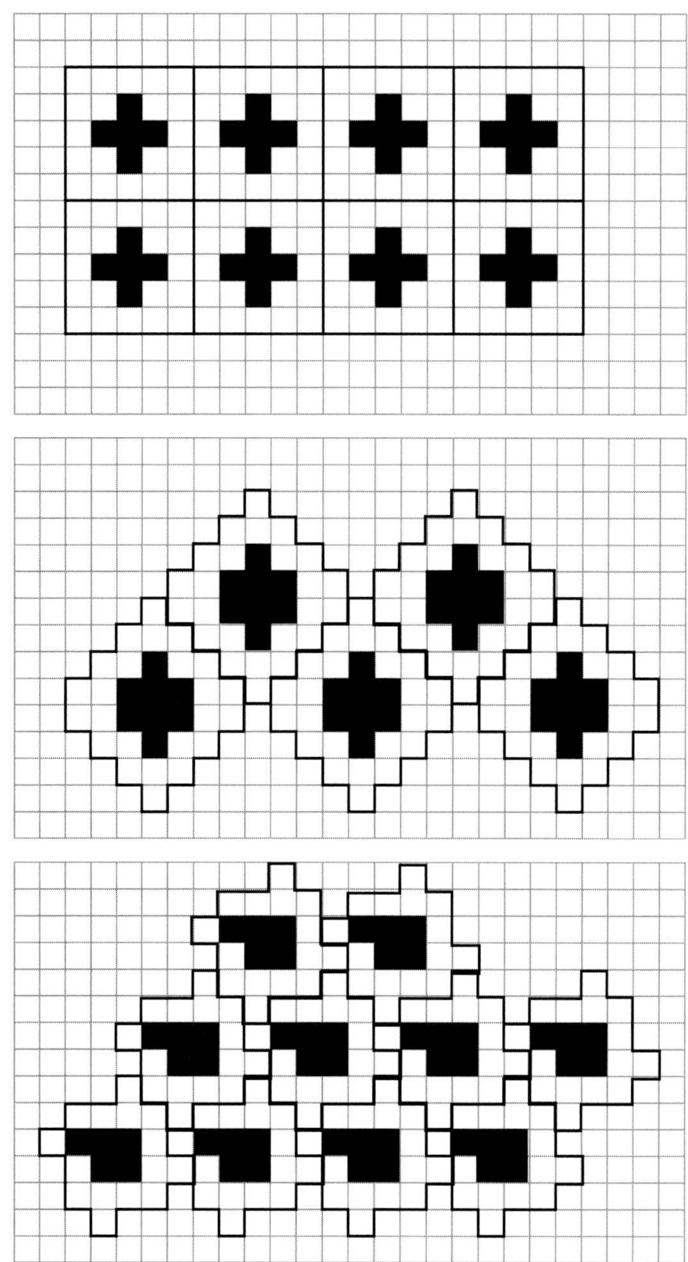

FIGURE 12-1
Halftone cells

TABLE 12-1
Requested versus actual frequency and angle at 300 dpi

	Requested Frequency			
	50 LPI		*60 LPI*	
Requested Angle	*Actual Angle*	*Actual Frequency*	*Actual Angle*	*Actual Frequency*
0	0	50.0	0	60.0
1	0	50.0	0	60.0
2	0	50.0	0	60.0
3	0	50.0	0	60.0
4	0	50.0	0	60.0
5	9	49.3	0	60.0
6	9	49.3	11	58.8
7	9	49.3	11	58.8
8	9	49.3	11	58.8
9	9	49.3	11	58.8
10	9	49.3	11	58.8
11	9	49.3	11	58.8
12	9	49.3	11	58.8
13	9	49.3	11	58.8
14	9	49.3	11	58.8
15	18	47.4	11	58.8
16	18	47.4	11	58.8
17	18	47.4	11	58.8
18	18	47.4	22	55.7
19	18	47.4	22	55.7
20	18	47.4	22	55.7
21	18	47.4	22	55.7
22	18	47.4	22	55.7
23	18	47.4	22	55.7
24	22	55.7	22	55.7
25	31	51.4	22	55.7
26	31	51.4	27	67.1
27	31	51.4	27	67.1
28	31	51.4	27	67.1
29	31	51.4	27	67.1
30	31	51.4	37	60.0
31	31	51.4	37	60.0
32	31	51.4	37	60.0
33	31	51.4	37	60.0
34	31	51.4	37	60.0
35	31	51.4	37	60.0
36	37	60.0	37	60.0
37	37	60.0	37	60.0
38	37	60.0	37	60.0
39	37	60.0	37	60.0
40	37	60.0	37	60.0
41	37	60.0	37	60.0
42	45	53.0	37	60.0
43	45	53.0	37	60.0
44	45	53.0	37	60.0
45	45	53.0	45	53.0

FIGURE 12-2
Cells and dot boundaries

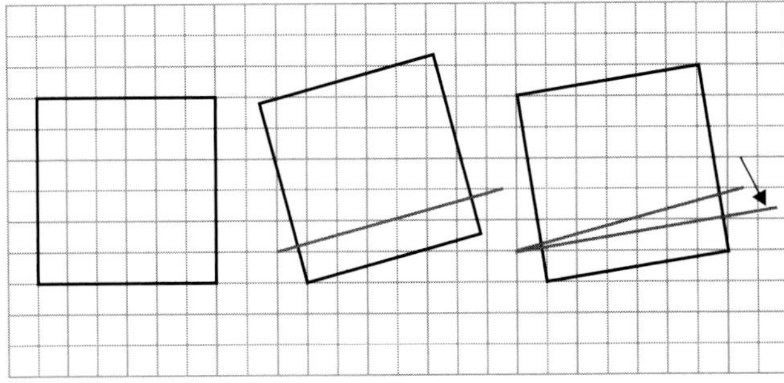

With fewer printer dots, there are fewer possible angles. The spots snap to the nearest printer dot.

FIGURE 12-3
Finer-resolution dots

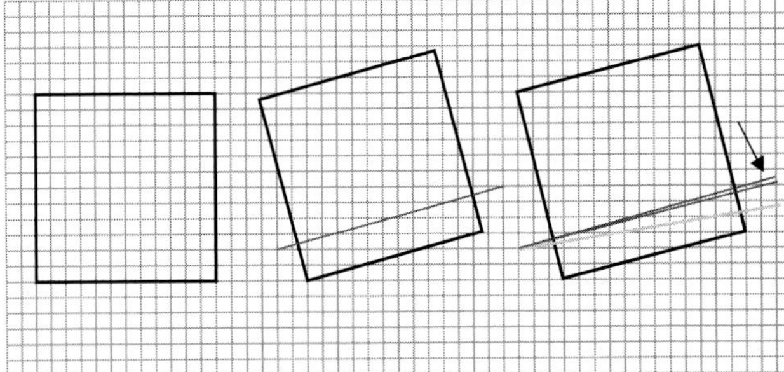

With a finer resolution, there are more dots to snap to, so more angles are possible.

scribing a rational tangent angle. In fact, any angle that you can get by using whole integers is rational. The digital halftoning that we've been describing works this way.

However, as we've seen in this chapter, many angles can't be generated this way; at least not at a particular screen frequency and output resolution. For example, there is no way to get a true 15-degree halftone screen from an imagesetter with rational screening. You can get close, but you can't achieve exactly 15 degrees because there are no integer values of dots that result in that angle.

Irrational tangent screening was developed and implemented on various RIP/imagesetter combinations to deal with this problem. In irrational screening, each halftone cell is not exactly the same size; some have more printer dots in them, some have fewer (see Figure 12-4). By fudging the process like this, you can achieve a halftone grid that is extremely close to the requested frequency and angle.

The problem with irrational screening is that it's incredibly processor-intensive (read: slow to print). Unlike rational screening, in

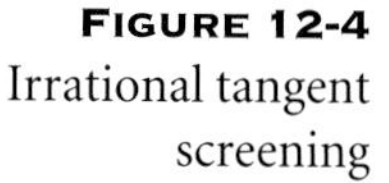

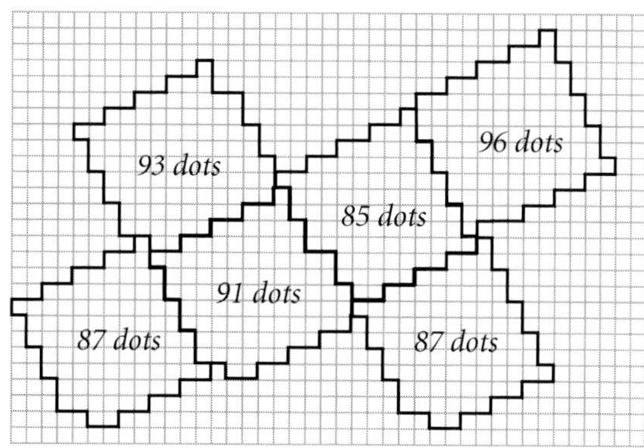

FIGURE 12-4
Irrational tangent screening

which each of the 256 possible spots is built and cached for fast access, with irrational screening, each and every spot must be calculated on the fly, as the image is being generated.

SUPERCELL SCREENING

If you take rational and irrational screening technologies and mix them together, you get an interesting hybrid called supercell screening. A *supercell* is a collection of smaller halftone cells, each of which may be made up of a different number of printer dots. So, while each of these small halftone cells may have different printer dots (much like irrational screening), the whole supercell collection is repeated and tiled throughout the halftone (more like typical, rational screening; see Figure 12-5).

The supercell can produce a more accurate screen/angle combination because it's much bigger than its constituent halftone cells, so there are more potential corner points to snap to. This sort of screening can be a great help in reducing or eradicating patterning; however, there's often a performance hit. It's faster than irrational screening, but not as fast as standard rational screening.

WHEN YOU NEED AN ANGLE

It probably seems, after reading this chapter, that the situation is almost hopeless. But that's no so. There are more and more ways to get really great, moiré-free color images. The supercell method is one

FIGURE 12-5
Supercell screening

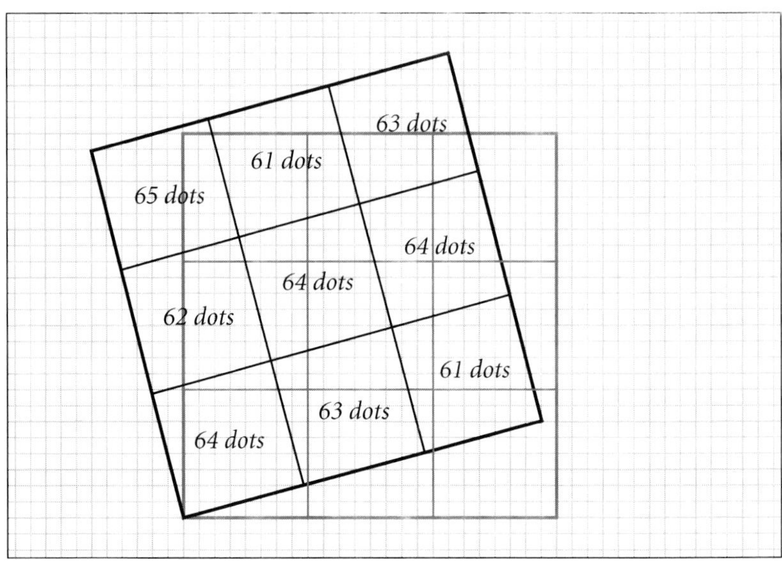

of them. Another one may be by simply working with th screen/angle combinations that your imagesetter manufacturer recommends.

Many lithographers are so caught up in the old way of doing things that if you show them a set of digitally-created screens at weird angles and frequencies (weird to them, at least), they'll say it won't work. But sometimes you know more than they do about what you can get, and about what you need.

13 CONTROLLING HALFTONE SCREENS

WHAT OVERRIDES WHAT

Whether or not you're a control freak, you're going to end up wanting to control your halftone screens at various stages of your work. You might want the picture of Aunt Jean to be a 45-degree, 40-lpi image, that photo of Uncle Izzy to be much coarser, and all the other tints on your page to have 133-lpi screens. So how do you go about controlling those screens, making sure the right screen comes out at the right time?

The answer is (you saw this coming, right?): it depends. Screening controls work differently with different combinations of hardware and software. There is a basic hierarchy we can set down, though, of which screening controls override which others—which controls predominate.

Figure 13-1 shows, in brief, the order in which screening controls override each other. Let's look at each item in order.

DEVICE'S DEFAULT SETTINGS. Every PostScript output device has a built-in default screen setting. That means that if you print a plain ol' job with no halftone screen settings in it, you get that device's defaults for the whole job. On imagesetters and other devices with hard disks, those defaults can be changed. (Be careful of the settings discussed below—especially driver settings for Windows users—you may be unintentionally getting halftone screens that you don't want because of the application or driver controls).

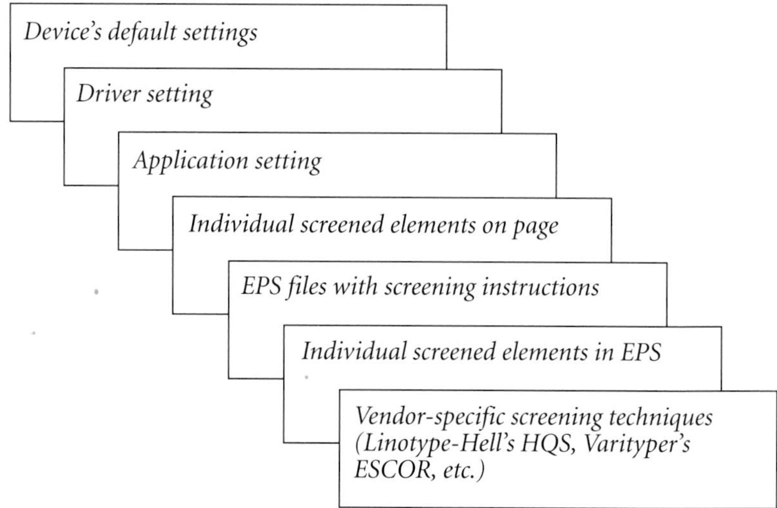

FIGURE 13-1 What screen settings override what

DRIVER SETTING. Printer drivers are the software modules that "drive" printers in the background; PostScript drivers actually write much or all of the PostScript code that gets sent to the printer. Some printer drivers (most notably the Windows 3.1 PostScript driver) let you control the halftone screen for the print job. In Windows, this control is in the Advanced Options dialog box (accessible via either the Setup button from the Print dialog box, or the Printer Setup command). This setting overrides the device's defaults.

APPLICATION SETTING. Many applications provide control over halftone screens for your print jobs. In QuarkXPress, this control is in Page Setup. In PageMaker, it's in the Print Color dialog box. In Photoshop, you get at it via the Screens button in the Page Setup dialog box. It could be in a variety of places depending on the application; the important thing to understand is that any screen information you set here overrides the driver and device default settings for the whole document.

INDIVIDUAL SCREENED ELEMENTS WITHIN PUBLICATIONS. In some applications—FreeHand and CorelDraw are examples—you can select individual objects (text or graphics) and set a screen for those objects. In others (such as PageMaker and QuarkXPress), you can apply screens to individual bitmapped images. These are called object-level settings, and they override the application-level settings, which apply to the rest of the job.

EPS FILES THAT INCLUDE SCREENING INSTRUCTIONS. Programs such as Aldus FreeHand, Adobe Photoshop, and QuarkXPress let you save pages as EPS files. These files are often saved with halftone screening information included. When you place this file in another program, those internal screening instructions override any screening controls set by the program you're printing from. The custom screening only affects the EPS that contains the instructions. The application, driver, or device defaults apply to the rest of the job.

INDIVIDUAL SCREENED ELEMENTS WITHIN EPS FILES. An EPS file that contains screening instructions can also include individual elements within the file that have *their own* screening instructions. For example, an EPS from Aldus FreeHand might have a gray box that has an object-level halftone screen applied to it. The EPS file's settings apply to the whole graphic, but are overridden for individual objects within the EPS that have their own settings.

VENDOR-SPECIFIC SCREENING TECHNIQUES. If you're printing to an imagesetter that uses a specialized screening technique such as Linotype-Hell's HQS or Tegra Varityper's ESCOR, you may not get the screen settings you expect. These techniques use screening "filters" which catch all screening instructions, and replace the frequency/angle combinations with the closest settings that are available in their optimized sets. These screening technologies can override all the screening you specify, whether it's in an object, an application, or a driver.

Who's On First

Trying to figure out what's controlling the screens in your print jobs can seem confusing. In most cases, however, you simply specify some halftone screen settings in the Print dialog box, and that setting applies to the whole job. Case closed. But, if you're not getting the screens you want, or if you want different screen settings for different parts of a job, this hierarchy of what overrides what should be a good roadmap for finding your way to the output you want.

Part 2
Scanning

14 Waltzing Through the Process

An Introduction to Scanning

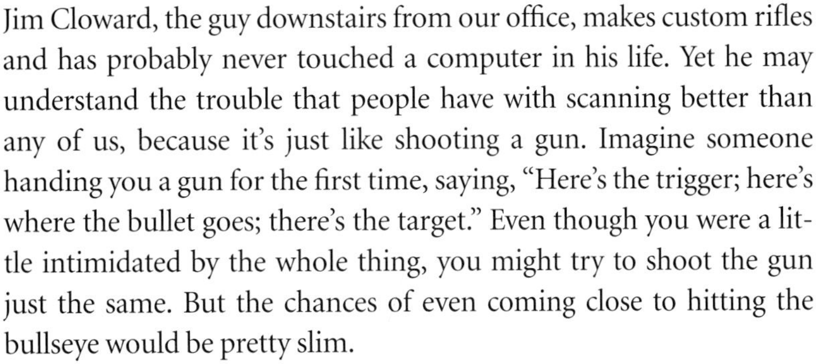

Jim Cloward, the guy downstairs from our office, makes custom rifles and has probably never touched a computer in his life. Yet he may understand the trouble that people have with scanning better than any of us, because it's just like shooting a gun. Imagine someone handing you a gun for the first time, saying, "Here's the trigger; here's where the bullet goes; there's the target." Even though you were a little intimidated by the whole thing, you might try to shoot the gun just the same. But the chances of even coming close to hitting the bullseye would be pretty slim.

This is essentially what has been happening in design firms and corporations for years. People come into an office one day to find a scanner on the desk with a manual on top of it. Seems easy enough: Just throw the artwork down, turn it on, point and shoot (as it were). However, chances are that the image they get in the computer isn't even close to the bullseye, and for the next two years they just "forget" to turn the scanner on in the morning.

If you're in the desktop publishing business long enough, you'll eventually find that you absolutely need to get an image from the "real world" into the computer. It might be a photograph, or a sketch, a page of text, or a picture from a home movie. You'll have to use a scanner of some sort. Don't be intimidated; we're here to tell you that it's not as complex as you might think.

Real World Scanning

Although we could talk all day about scanning, there are really only five basic elements of scanning that you need to think about.

- Type of scanner
- Scanning mode
- Scanning resolution
- Adjusting the scanned image
- Printing the image

When scanning an image, you have to make decisions in each of these areas. What kind of scanner can I use? What resolution should I scan at? How am I going to output the image later on? The problem that most people have in scanning is that they don't understand the interconnectedness of each of these decisions. The scanning resolution depends on the output method, the type of scanner depends on what kind of art you have, the types of adjustments you do to the image depend on the manner in which you scan the image, and so on.

Let's take a very brief look at each of these areas, outlining their key issues. We'll go into more detail in each of these areas in the subsequent chapters.

Type of Scanner

The type of scanner that you use is determined almost entirely by the image you wish to capture (and your budget, of course). If you're trying to scan someone's face without photographing it first, you have to use a device such as a video camera or a still digital camera. If you've got a color transparency, you'll have to use a scanner that takes transparent art, such as a slide or a drum scanner. For reflective art, flatbed scanners work fine, but drum scanners are better.

The second consideration is the quality of the scan you're trying to achieve. A scan from a drum scanner has much higher quality than from a flatbed scanner. And a hand-held scanner can rarely achieve the results of a flatbed scanner (again, it often depends on what you're scanning).

We'll take a look at the various types of scanners and what their strong and weak points are in Chapter 15, *Scanners*.

SCANNING MODE

One key to creating a good scan is picking the correct scanning mode. Typically, this means choosing among line-art, dithered, grayscale, and color. There's almost no reason to scan a black-and-white, line-art image in 24-bit color mode. Similarly, there's almost no reason to scan a beautiful photograph in high-contrast black and white.

The mode in which you choose to scan is dependent on what you're scanning, what you plan on doing with the scan on the computer, and how you plan to output it. If you're going to retouch an image in a program like Photoshop, you'll avoid dithered images. If you're trying to save space on your hard disk, you might consider a line-art scan.

If the words dithered, grayscale, or line art don't mean much to you, you probably want to take a look through the earlier chapters on halftoning. We'll take a further look at scanning modes in Chapter 15, *Scanners*, and Chapter 16, *Scanned Images*.

SCANNING RESOLUTION

If you have a 300-spi scanner, you should scan at 300 spi, right? Wrong. To pick the correct scanning resolution, you need to think about file size, output resolution, screen frequency (if it's a grayscale or color scan), and the size of the image—the original image and the final image that you output. And all of these items interrelate. (By the way, we'll discuss the difference between the terms "dpi" and "spi" in the next chapter.)

We talk about these issues in Chapter 18, *Image Resolution* and Chapter 19, *Choosing Resolution*.

ADJUSTING THE SCANNED IMAGE

Once the artwork is scanned in using a particular scanner in a particular mode and resolution, you have to save it as a file on a disk. And once it's on disk, you can change its tonal levels (contrast and brightness), manipulate the picture (move a tree in the background, delete a person from the scene, and so on), and make it sharper. We'll talk about these issues in Chapter 17, *File Formats,* Chapter 20, *Tonal Correction,* Chapter 21, *A Sharper Image.*

You can perform all of these functions—plus create lots of other effects—in programs like Adobe Photoshop or Aldus PhotoStyler, which we cover in Chapter 27, *Image Applications.*

PRINTING THE IMAGE

The last area of scanning you need to think about is how you'll get the image out onto paper or film. This includes halftoning to black-and-white imagesetters (which we've been talking about in the book up until now), and output to color printers and film recorders.

But the issue is larger than just the final output medium; it also includes screen frequency of halftones, file size, file type, and a host of other factors depending on the software that you're using (QuarkXPress versus PageMaker versus Photoshop, and so on). With all these items to think about for what seemed like a simple scan, it's amazing that anyone ever gets them right! Fortunately, the bark is worse than the bite; we're confident that after reading this section you'll have all the necessary information to create great scans.

15 Scanners

What You Need and What You Get

If you've never been to a big electronic publishing exhibition like The Seybold Conference, you probably don't understand why David Cronenberg does what he does. Mr. Cronenberg made a movie some years ago called *Scanners*, about people who could control someone's mind so strongly that occasionally it would make their head explode (we even considered calling this book *Scanners*, and using that clip for the cover). Now that's not a very pleasant thought, but big exhibitions and conferences bring that image to mind. Dozens and sometimes hundreds of manufacturers combining their marketing muscles in an attempt to control your mind has been known to produce a very similar effect (although taking an Advil usually cures the problem).

One of the biggest causes of the "exploding head" syndrome is looking at scanner manufacturers' marketing materials. There are so many different types of scanners, and all of them are supposed to be the best. In this chapter we want to explain a little about various sorts of scanners, what they can and cannot do, and what makes a good scanner.

There are four major factors to consider in evaluating scanners: resolution, bit depth, software, and scanner type.

Resolution. How many samples per inch (spi) can the scanner capture? This seems simple, but isn't always. There's optical resolution and interpolated resolution, and resolution needs vary depending on the size of the art you're scanning.

Bit depth and dynamic range. How many bits per sample does the scanner capture? Again, a simple number doesn't tell the whole story. All eight-bit scanners are not created equal. The real question is how well the scanner can differentiate subtle differences in tone across the range from black to white.

Software. What capabilities are built into the scanning software? Does it just capture an image and drop it on disk, or can it apply tonal correction and sharpening as it scans? And how good are those tools? Does the scanner only work with a standalone application, or can you access it via plug-in modules from within other programs? This has become less of an issue because of Hewlett-Packard's TWAIN system, explained later.

Scanner type. Is the scanner a flatbed, sheetfed, slide scanner, video capture board, digital camera, drum scanner, or hand scanner? The type you use determines the types of images you can capture, and different types have different strengths and weaknesses in the three areas mentioned above.

Resolution

It seems like it should be easy to compare different scanners' resolution capabilities. One scans at 300 spi (be aware that manufacturers usually specify resolution in dpi, but spi—or samples per inch—is really a more accurate term), and another scans at 600 spi, so the 600-spi scanner is better. Right?

Not necessarily. Most manufacturers list their scanners' *optical resolution* in their marketing materials. That's the maximum number of samples per inch that the scanner can actually discern. Some manufacturers, however, list the highest *interpolated resolution* that their scanner/software combination can provide. You can read Chapter 18, *Image Resolution*, to find out what interpolated resolution means, but we can tell you right here that listing interpolated resolution rather than optical resolution is deceptive. If a scanner manufacturer claims more than 300 spi for true resolution, check it out carefully. Make sure to determine a scanner's true optical resolution before buying.

Also bear in mind the effects of scaling on resolution. You need more resolution with some types of scanners than with others. A slide scanner that scans at 1,000 spi sounds nifty compared to a 300-spi flatbed, but when you consider that you often need to enlarge scans of slides to 400 percent (or even 800 percent), 1,000 spi doesn't sound so impressive.

Bit Depth and Dynamic Range

Bit depth determines how many grays or colors a scanner can discern. A one-bit-per-sample (or bilevel) scanner can only capture two gray levels—black and white. Four-bit scanners (there aren't many around any more) can discern 16 gray levels. Eight-bit scanners can capture 256 gray levels (there are a few eight-bit color scanners around, mostly of the hand-scanner variety). 24-bit scanners capture more than 16 million different colors, and there are scanners that capture even more information than that.

The fact of the matter is, however, that not all eight-bit and 24-bit scanners can capture the same amount of information. If you scan the same image with two different eight-bit flatbeds, for instance, one may do an excellent job of discerning subtle differences in gray levels, showing a lot of detail in shadow areas. The other will give you a scan where all the shadow areas are a uniform black or dark gray (see Figure 15-1). Compared to the first scanner, the second scanner has poor *dynamic range.*

The problem is noise. Just as the static—or noise—on a radio can hide the actual radio signal, the noise from a bad scanner can obliterate the useful information that it captures. Another way of saying this: the first six bits from most scanners are accurate (the scanner can accurately distinguish between 64 different gray levels). The seventh bit is less accurate (it has some trouble distinguishing

FIGURE 15-1 Good and bad dynamic range

Good dynamic range

Poor dynamic range

Notice how in the image on the right, the highlight areas—such as snowbanks—are washed out to white, lacking in detail. The shadow areas, while displaying reasonable detail due to post-scan tonal correction, are still missing detail compared to the image on the left. Even with the best of tonal correction tools, it's impossible to bring out both highlight and shadow detail in an image that is scanned originally with poor dynamic range.

between the next 64 gray levels), and the eighth bit is even less reliable (the top 128 gray levels are even more mixed—farther from what their actual values should be). Again, some scanners are far better than others, but no eight-bit scanner can accurately distinguish between a full 256 gray levels.

The same problem arises with 24-bit scanners, since they actually make three eight-bit scans—sometimes in separate passes—one each for the red, green, and blue channels.

CCD SENSITIVITY

The problem of inaccurate data is exacerbated by a simple fact about the photosensitive elements in most desktop scanners. The elements, called charge-coupled devices, or CCDs, have a linear response, while the transmission or reflection of film or paper is logarithmic. The result is that CCDs are unable to discern as many different gray levels in shadow areas as they can in highlights (the reverse is true when scanning negatives).

The result of this linear-versus-logarithmic dilemma is poor detail in shadow areas. Adjacent sample points that should be different values, aren't. And those differences are the stuff that detail is made of. If

Process Color

Cyan, 15 degrees

Magenta, 75 degrees

Yellow, 90 degrees

Black, 45 degrees

Duotones

Cyan plate

Black plate

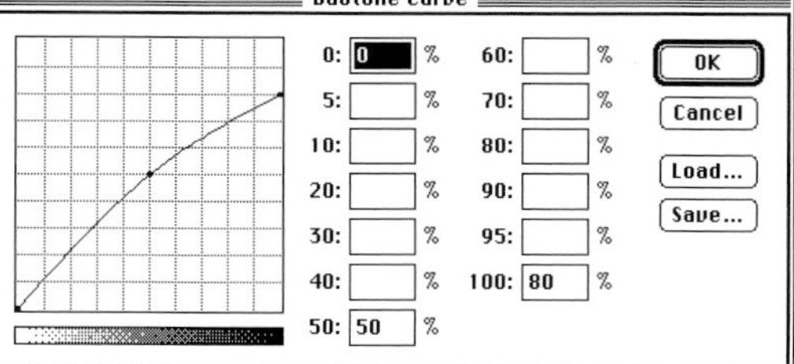

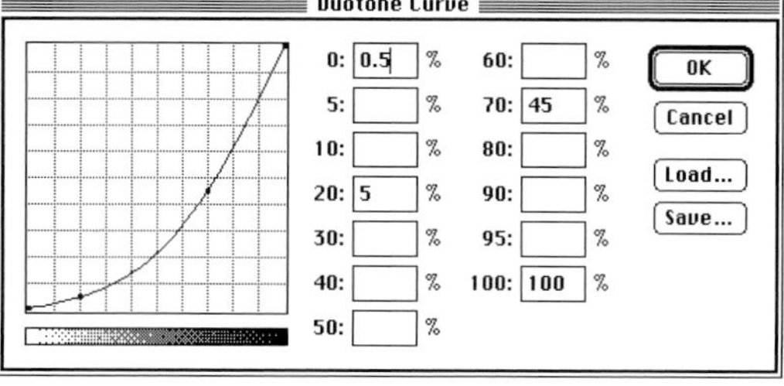

Rosettes

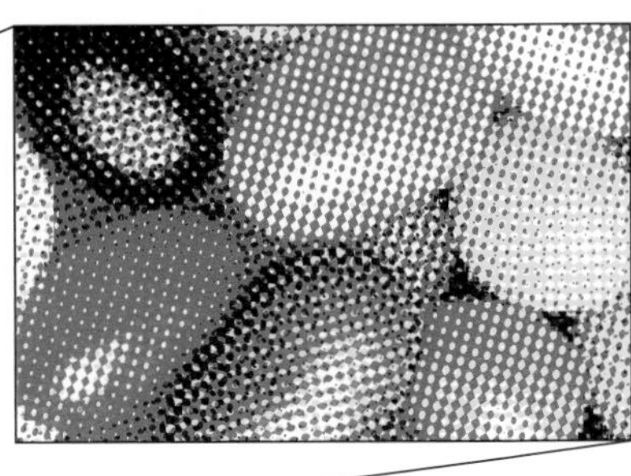

Moiré Patterns

Standard process color angles

Black plate rotated 10 degrees

Photo CD

Base resolution (512×768)

*Base*16 resolution (2048×3072)*

Both images were downsampled in Photoshop from their original resolution to 200 spi.

Comparison Scans

225-spi image from a drum scanner

225-spi image from a Photo CD

225-spi image from a drum scanner

225-spi image from a flatbed scanner

JPEG Comparisons

Original picture
575K

Fair compression
252K

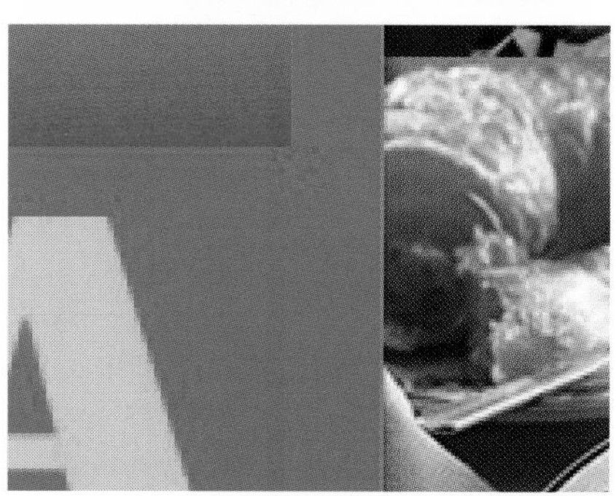

Good compression
62K

Excellent compression
17K

100% *250%*

you try to emphasize the details in shadow areas with tonal correction—accentuating the differences between adjacent samples—you lose detail in the highlights: either they wash out to white or they all become too similar to display much detail.

Manufacturers rarely, if ever, list dynamic range as a specification in their marketing materials, and even if they did, it would be hard to rely on because different testing methods for dynamic range can result in widely varying figures. Our only advice is to read multiple magazine reviews before deciding on purchasing a scanner. And don't be afraid to read a Macintosh review if you're buying a PC scanner (or vice versa). Most scanners work with both, and the testing results are equally illuminating on either platform.

INCREASING DYNAMIC RANGE

Some scanners on the market employ methods to increase the dynamic range of the scans they capture. Most of the scanners that employ these methods (with some exceptions) are slide scanners, drum scanners, and other units designed for quality graphic-arts work. Techniques include the following.

CAPTURING EXTRA BITS. Many scanners capture more than eight bits of data (or more than eight bits per color channel). Then they resolve the data down to eight good, accurate bits of information per sample point. This is really the best technique for capturing high-quality scans.

Some of these scanners also allow you to apply tonal correction and sharpening to the scan before downsampling, minimizing the data loss that's inevitable with these image correction techniques.

Leaf Systems has proposed a new standard for color scanned image files. In this new standard, scans would include 48 bits of RGB information (16 bits per color channel). While this might seem like overkill, it's what's necessary to accurately capture the dynamic range of film. And it allows for tonal correction with minimal loss of data. Once you're done with all image modifications, the file is sampled down to 24 accurate bits.

ADJUSTING EXPOSURE TIME. Several scanners (almost all slide scanners) can adjust the amount of time that each scanning element is exposed to light. Most figure the exposure automatically, doing a prescan then adjusting exposure based on the density of the image. By increasing the exposure time, you increase the amount of information to be interpreted and converted to digital form resulting in more accurate sampling. The trade-off is speed: longer exposure times make for slower scans.

CORRECTING ANALOG DATA. A couple of slide scanners on the market can apply tonal correction to analog sample data, before it's converted to digital form. Again, this minimizes the data loss associated with image correction.

ADAPTIVE ANALOG-TO-DIGITAL CONVERTERS. There are a few scanners available that include adaptive analog-to-digital converters (the devices that convert the sampled information into digital form). If the image you're scanning doesn't encompass the whole tonal range from black to white (it's a very flat, or very light, or very dark image), adaptive A/D converters can use the CCD's full ability to discern tonal differences, but apply it to a narrower range of values, resulting in better dynamic range. The hardware, in cooperation with the scanning software (the Apple Color OneScanner with Ofoto 2.0, for instance), handles this automatically.

SOFTWARE

There are three main things you should consider in scanning software (whether it comes with the scanner or you buy it separately): tonal correction, sharpening, and plug-in modules.

TONAL CORRECTION. If your scanning software allows you to apply tonal correction as you scan, you can save a lot of time when producing multiple scans. Without this capability, you need to scan, save, open the picture in an image-manipulation program, correct it, then save again.

The tonal correction that is offered, of course, has to be *good* tonal correction. As we point out in Chapter 20, *Tonal Correction*, linear correction (usually consisting of contrast and brightness adjustments) is essentially worthless. You need good tools for nonlinear correction, and (perhaps) for closed-loop calibration.

SHARPENING. It's often nice to be able to sharpen the scans as you capture them, especially when you're under tight deadlines. Again, it avoids the "scan/save/open/adjust/save" routine. And, again, you need to be sure that the software's sharpening routine is up to snuff. Sharpening is the last manipulation you want to perform on a scan, however, so if you sharpen while scanning, you can't make any further changes. Run some tests to your final output device to make sure that the image quality with your scanning software's sharpening is equal to the sharpening available in image manipulation packages.

ACCESS FROM OTHER PROGRAMS. Almost every scanner comes with a standalone application for controlling the scanner and capturing scans, but it's also convenient if you can scan images from within other applications. If your scanning application doesn't offer (good) tonal correction and sharpening, scanning directly into Photoshop or the like can partially short-circuit the multiple-save routine.

Most scanners come with plug-in acquire modules for Photoshop (and various other programs can use those Photoshop plug-ins), so you can scan straight into the program. Also, many scanners now support the TWAIN standard (believe it or not, this stands for "Technology Without an Important Name"), so any application that supports TWAIN can access the scanner.

SCANNER TYPES

As we mentioned in the last chapter, the sort of scanner you need at any given time depends mostly on the sort of image you're trying to get into your computer. Different scanner types also tend to perform differently in the areas discussed above: resolution, dynamic range, and software. Let's take a quick overview of the various types of scanners, considering them for types of artwork and quality they support.

Flatbed scanners. The most common scanner in people's offices today is the flatbed scanner. This usually looks like some weird cross between a photocopier and a bizarre packing carton for an alien computer. The idea behind flatbed scanners is that you can put flat artwork down on a sheet of glass, and a scanning mechanism moves under it, capturing the image. Flatbed scanners are generally set up for reflective art (like pictures, pages, and so on), but they can often be fitted with attachments that enable them to scan transparent art (slides, transparencies, and so on).

Although flatbed scanners are pretty solid, they're often cheaply made because there's a great pressure to market scanners inexpensively to businesses. Problems crop up with blurry or jittery scans when the scanning mechanism itself doesn't move smoothly or is on a shaky surface. Some offer impressive dynamic range, however, and some even offer bit depths beyond eight or 24 bits.

Sheetfed scanners. Sheetfed scanners are less common these days. Somewhere between a fax machine and a form-feed printer, these scanners generally pull or push an image past a stationary scanning mechanism. The biggest problem here is that sheetfed scanners can only scan artwork that is flexible enough to be pulled through the pasta-making innards. That means pictures mounted on artboards and books are out. And often smaller images have to be mounted on a larger piece of paper before they can successfully navigate through the scanner. It's also extremely difficult to feed the art accurately so that horizontal and vertical lines end up horizontal and vertical.

Sheetfed scanners are used almost exclusively for optical character recognition (OCR) work these days, though they're also useful (if you can get one cheap enough) for grabbing low-resolution FPO (for-position-only) images for placement in your layouts.

SLIDE SCANNERS. Where the last two scanner types are primarily used for scanning reflective art, slide scanners are used exclusively for transparent art. Also note that the resolution of slide scanners is often much higher than flatbed or sheetfed scanners; but that's only because they're typically scanning much smaller objects (like 35mm film) which need to be enlarged. Slide scanners also tend to deliver better dynamic range than flatbeds, since they're typically designed for the graphic-arts market, rather than for general business use. Most slide scanners capture at least 10 bits per channel; some capture as much as 16 bits.

DRUM SCANNERS. The highest quality scans usually come from drum scanners. These devices (starting at around $35,000 and going up into the stratosphere) use photo-multiplier tubes (PMTs) instead of CCDs to capture gray levels, resulting in greatly superior dynamic range. With a drum scanner, you tape the artwork down onto a cylinder (the drum). The scanner spins the drum very quickly while moving the PMTs over the artwork.

Drum scanners are often incorporated into complete image-scanning and processing systems (though you can just plug some of them into your Mac these days), so more often than not you'll send images to a scanning service rather than using the device yourself. This is currently the method of choice for quality-critical work.

HANDHELD. On the other end of the scanner spectrum are handheld scanners. These are tiny reflective-art scanners that you physically move across an image. They're useful for some things—quick scans of text for OCR or low-quality scans for FPO work—but their quality isn't good enough for most graphic arts work. Nonetheless, they're popular for low-end scanning jobs because they are so inexpensive. There are even some color hand scanners on the market, though they face the same quality problems.

VIDEO. We didn't think that video scanning was a reasonable way to capture images (everything we'd seen looked terrible) until a friend showed us the poster he produced that included 50 video captures. Now we know it's possible to get good images from a video tape or video camera; you just have to work hard to do it.

Video capture setups are typically in the form of a board that you plug into your computer. You feed the video in, and then control the capture of a particular frame of the video through software. The problems of converting a frame of NTSC or PAL video to an RGB color scan are not slight. However, video capture is becoming more and more common, especially for creating special effects or for images that will be converted back to the video medium.

If you're planning on capturing video images from tape, consider using the Hi-8 format instead of VHS or the like, because Hi-8 stores more and better information.

Digital cameras. One of the newest image-capture technologies is the digital camera. As we write this (Fall 1993), there are very few cameras available, and those are quite expensive. Most of these digital cameras actually consist of a digital scanner added to the back of an existing camera. For example, the Leaf Digital Studio Camera system adds a device to the back of a Hasselblad or a Mamiya studio camera.

The camera acts just as if you were using regular film, but the image is exposed onto a photosensitive chip that contains millions of sensors—usually CCDs. The information is converted on the fly into a digital picture and passed to the computer to save on a hard drive (or it may be stored on the camera's hard drive).

Whereas some digital cameras, like the Leaf system, are for studio use only, others (such as the Kodak Digital Camera System, which is built around a Nikon 35mm camera) are portable and battery-operated. Although these smaller systems don't presently provide a very high-quality picture—they have limited dynamic range and image size—they may be useful for some applications such as real-estate photos and some photojournalism. Note, however, that most digital cameras presently have a several-second waiting period while the image is captured and stored on disk. Sometimes this can take up to 10 minutes, making it impossible to capture any live action quickly. There are exceptions, such as the Kodak DCS 100, which can capture 25 images .2 seconds apart, with ensuing images 2.5 seconds apart.

Hold on to Your Hat

Obviously, there are a lot of issues to think about when considering scanning something. If your head isn't exploding now, we just don't know what else we can do—except direct you to the next chapter, where we get into the details of scanned images. But don't fear; the more you learn about this stuff, the easier it all gets.

16 Scanned Images

Anatomy of a Bitmap

There's an amazing phenomenon that happens whenever someone is talking about desktop publishing: every time the word "bitmap" is mentioned in conversation, people's eyes glaze over. Really; try it sometime. You can say, "Scan that photograph, drop it in the page-layout program, and print the halftone on the imagesetter," and everyone will nod their heads and smile. Then say, "The resolution of the bitmapped graphic doubles when you reduce the image's size to 50 percent," and your audience will be asleep before you finish the sentence.

Bitmapped images do not make for exciting conversations. But it is a topic that is incredibly important to know about when working with scans. In the next few chapters we're going to cover everything you need to know about bitmapped images. We promise it isn't as confusing—or as boring—as it sometimes seems.

In this chapter we describe the characteristics of bitmapped graphics—the attributes that define them—and how those characteristics relate to file size.

What is a Bitmap?

To begin with, many people want to know what a bitmap really is, and why it's called a bitmap. It's actually a pretty descriptive term, once you understand the underlying concepts.

A bitmap is just that—a map of *bits* (pieces of computer information) describing all the points in a rectangular grid of dots. In plain English, a bitmapped file might read, "The point at coordinates 1,1 is black; the point at coordinates 1,2 is white; the point at coordinate 1,3 is black," and so on. It's essentially drawing a map of the bits in the graphic.

SAMPLE POINTS

There's one other term we need to discuss here in relation to bitmaps. When we talk about points in a bitmapped graphic, we call them *samples* or *sample points*. We use this term because a scanner *samples* an image—checking what color or gray value it finds—every 1⁄300 of an inch, every 1⁄100 of an inch, or whatever. So we talk about the resolution of bitmaps in *samples per inch*, or spi.

Samples are not the same thing as dots (the things that laser printers and imagesetters make) or spots (the elements of a halftone) or pixels (the "picture elements" on a computer screen), so we prefer to use a distinct term for them.

Characteristics of Bitmapped Graphics

Every bitmapped graphic has four basic characteristics: dimension, resolution, bit depth, and color model. People use these terms all the time without really knowing what they mean, and then make really poor decisions based on a faulty understanding. So let's look at what these words refer to, and why they're important.

DIMENSIONS

Bitmapped images are always big rectangular grids. Like checkerboards or chessboards or parquet floors in your kitchen, these big grids are made of little squares. And the one thing you can always say about grids is that they have dimensions (see Figure 16-1). A

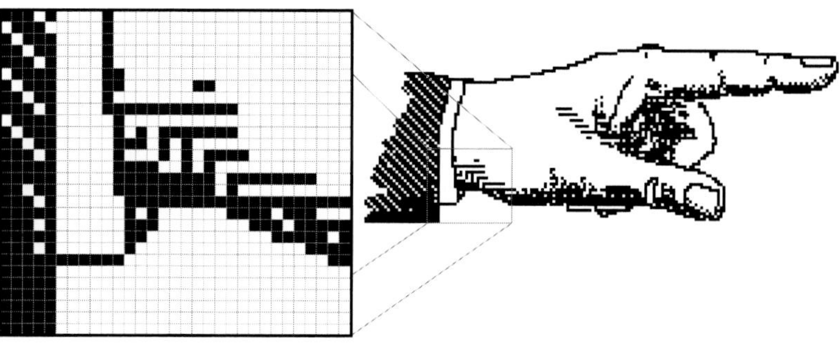

FIGURE 16-1 Dimensions of a bitmapped image

chessboard is always eight squares by eight squares. The grid of pixels that makes up your computer screen might be 640 by 480.

Be aware that just because you know the dimensions of a grid, you don't necessarily know how big it is. You don't know how close together the samples are. We're talking number of samples, not measurements. That's where resolution comes in.

RESOLUTION

The word "resolution" is so misunderstood and overused that we've devoted two chapters to it later in the book (Chapter 18, *Image Resolution* and Chapter 19, *Choosing Resolution*). The resolution of a bitmapped image is the number of samples in each unit of measurement. If we're talking in inches, then we talk about the number of samples per inch (spi).

If your bitmapped image has 72 dots per inch, and it's 72 dots long on each side, then you know that it's an inch long on each side. We can take the same bitmapped image and change it to 36 dots per inch (change its resolution), and suddenly the image is two inches on each side (same number of dots, but each one is twice as big; see Figure 16-2).

You can also look at bitmap resolution in another way: If you know the size of an image and its resolution, you can figure out its dimensions. When you scan a picture that is three inches on each side at 100 samples per inch, you know that the bitmapped image has 300 samples on each side (100 per inch). If you scan it at 300 samples per inch, the dimensions shoot up to 900 dots on each side.

FIGURE 16-2 Changing bitmap resolution

Because each dot takes up a little room on disk, the more dots, the larger the disk file is.

Increasing the resolution of a bitmapped image increases file size geometrically. So if you double an image's resolution, you increase file size by a factor of four. Increase to three times resolution, and file size increases by a factor of nine. For more on resolution, see Chapter 18, *Image Resolution*, and Chapter 19, *Choosing Resolution.*

BIT DEPTH

Each sample in a bitmapped image can be black, white, gray, or a color; the key is the number of bits used to describe it. A *bit* is the smallest piece of computer information; it's either a zero or a one. A zero means "white," a one means "black." Therefore, a sample that is defined using one bit of information can only be black or white—zero or one.

If you have two bits of information describing a sample, there are four possible combinations (00, 01, 10, and 11), hence four possible colors or gray levels (see Figure 16-3). Eight bits of information gives you 256 levels of gray; 24 bits of information results in over 16

FIGURE 16-3
Bits per sample

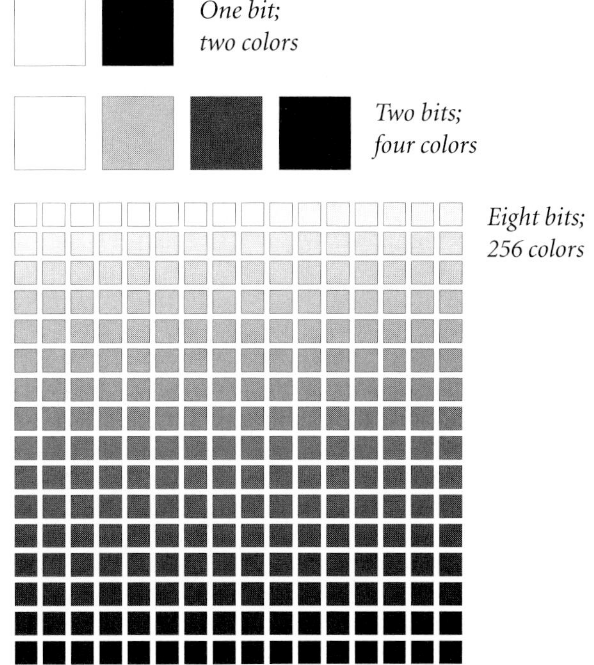

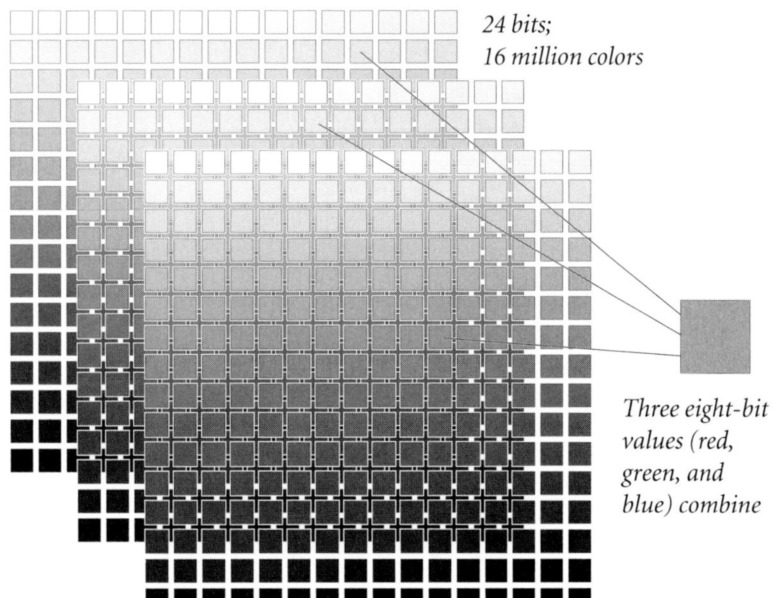

million possible colors (with 24-bit images, each sample actually has three eight-bit values—one each for red, green, and blue).

The number of bits is called the image's bit depth. We call a one-bit image a *flat* bitmap, also called a bilevel image. A *deep* bitmap is any image that has more than one bit describing each sample (see Figure 16-4).

Bit depth also affects the image's file size. An eight-bit image is eight times the size of a similar one-bit image: it uses eight bits to describe each dot instead of one. A 24-bit image is 24 times as big, and so on. Increasing bit depth increases file size arithmetically.

FIGURE 16-4
A "deep" bitmap

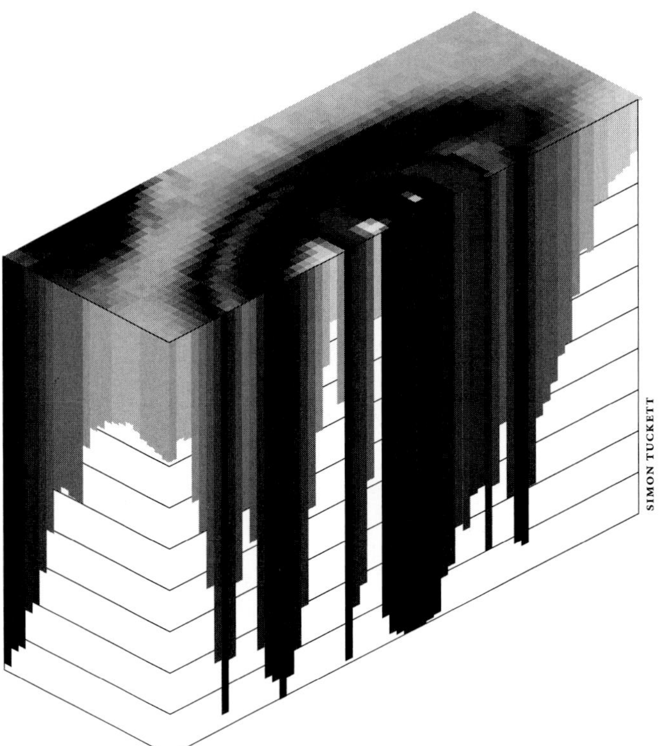

COLOR MODEL

The final attribute of a bitmapped image is the color model, which really only applies to color images. The most common color model is RGB (red, green, and blue). Some bitmaps are described using the CMYK color model, defining values for each of the four process colors: cyan, magenta, yellow, and black. While this results in 32 bits per sample, it doesn't actually add more colors; it's just a different method of describing the colors—a method that's useful for process-color work. For more on CMYK bitmaps, see the discussions of the TIFF and EPS formats in Chapter 17, *File Formats.*

DEVICE-INDEPENDENT (CIE) COLOR. A new class of color models has emerged of late, designed to work with the color-management systems now available that improve color correspondence between your screen, color printouts, and final printed output. These *device-independent* or *perceptually based* models don't describe a color by the components that make it up (RGB or CMYK, for instance). They describe *what a color looks like.* All of them are based, more or less, on the standards defined by the Commission Internationale de l'Eclairage (CIE) in 1931.

The problem with RGB and CMYK color models is that a given RGB (or CMYK) specification doesn't really describe a color. A color specified using RGB may look totally different on your monitor than it does on color-printer output. Perceptually-based color definitions describe what a color looks like; it's up to your color-management software to decide what RGB or CMYK values are needed to create that color on a given device.

Color-management systems and their associated color models are still in their infancy. It remains to be seen which will predominate in the market, and how different systems will work together.

INDEXED COLOR. Indexed color is a method for producing eight-bit, 256-color files. Indexed-color bitmaps use a table of 256 colors, chosen from the full 24-bit palette. A given sample's color is defined by reference to the table: "this sample is color number 123, this sample is color number 81," and so on.

While indexed color can save disk space (it only requires eight bits per sample point, rather than the full 24), it only gives you 256 colors (albeit optimized for that image), and sometimes the tables don't match between applications, so you get weird color changes. You might want to convert indexed-color bitmaps to true RGB, though you won't improve the image any in the process—you're still only getting 256 colors.

Unlike resolution and bit depth, the color model of a bitmapped image doesn't necessarily have a direct bearing on file size. However, if you change the color model, the computer will most likely change the image's bit depth, changing file size. For example, if you convert a CMYK image to a grayscale image, the computer internally changes the bit depth from 32 bits (8 bits for each of the four channels) to 8 bits. This cuts the file size by 75 percent. Similarly, if you change a flat, bilevel bitmap into an RGB bitmap, the computer has to add 23 bits to each pixel, making the file size 24 times as large.

Bitmaps and File Size

You've no doubt noticed that throughout this chapter we've commented on how different bitmap attributes affect file size. We've done so because it's a really important topic. Big files are hard to work with, hard to print, and hard to transfer. So it's well worth reducing file size when you can.

To encapsulate what we've said above, here's a rundown of the characteristics of bitmapped images that affect file size.

Dimensions and Resolution. Increasing the number of samples in a bitmap (by increasing either the dimensions or resolution) increases file size geometrically: 200-spi bitmaps are four times as big as 100-spi bitmaps, and 300-spi images are nine times larger.

Bit depth. Increasing bit depth increases file size arithmetically. A 24-bit image is three times as large as an 8-bit image.

Color model. Color model doesn't necessarily increase file size, but going from RGB (24-bit) to CMYK (32-bit) color models does because it alters bit depth.

FIGURING FILE SIZE

Now that you know the factors that affect the size of bitmaps, it's a simple matter to calculate file size using the following formula.

File size (kilobytes) = Resolution^2 × Width × Height × Bits per sample ÷ 8,192

This formula works because 8,192 is the number of bits in a kilobyte. For example, if you have a four-by-five-inch one-bit image at 300 spi, you know that the file size is 220K. A 24-bit image of the same size would be 5,273K (which is just about five megabytes).

FILE FORMAT

One other factor—file format—can also affect the file size of bitmapped image files (we discuss these in detail in Chapter 17, *File Formats*). Storing an image as a TIFF file creates a different file size than storing it as a PICT or EPS file, even if the dimensions, resolution, and bit depth are the same. Why? Because the EPS file typically includes a preview image over and above the image stored in the file. PICT and TIFF files often include built-in compression (see Chapter 22, *Getting Small,* for more on compression), making files smaller. And other file formats include similar additions or compression methods to suit different needs.

Plus, when you're saving bitmapped images in EPS format, there's one other consideration—binary versus ASCII encoding. We won't bother you with an in-depth description of PostScript's *readstring* and *readhexstring* operators. We'll just tell you that binary-encoded bitmaps are half the size of ASCII-encoded bitmaps. ASCII encoding is often less problematic, however, when you're passing files around through different kinds of networks and computing platforms.

IMPLICATIONS OF FILE SIZE

Why should you care about file size? Because big files are harder to store, transfer, manipulate, and—especially—print. This section describes the problems posed by large bitmapped image files. For advice on reducing file size, see Chapter 18, *Choosing Resolution.*

STORAGE. Hard disks come cheap these days, but they're still not free (we can't imagine why!). Smaller files save you money because you don't have to buy new hardware. And if you want to give images to a colleague or service bureau, you have to fit them on some kind of disk or tape. As David's grandmother always says, "smaller is better."

MEMORY. You often don't need a lot of RAM to open an image (and you can usually place images on pages regardless of how much RAM you have), but speed suffers radically. If you can fit a whole image in memory, you can manipulate it between five and 50 times faster than if you have to work off a disk.

Some scanning and image manipulation packages *require* that the whole image fit in memory. With these packages, your only choices are to reduce file size or buy RAM.

PRINTING. The biggest problem with big files is printing. You have to transmit the files over some kind of cable (serial, parallel, Ethernet, AppleTalk, whatever), and the printer or imagesetter has to digest all the information. At best, big files make you wait for output. At worst, you incur additional imagesetting charges for slow output—or the files won't print at all. Keep files small to avoid all these problems.

What's In a Bitmap

Now that we've covered the attributes that define what's in a bitmap, in the next chapter, *File Formats*, we'll move on to discuss the different containers available for saving that information to disk and passing it between programs.

17 File Formats

Which One and Why

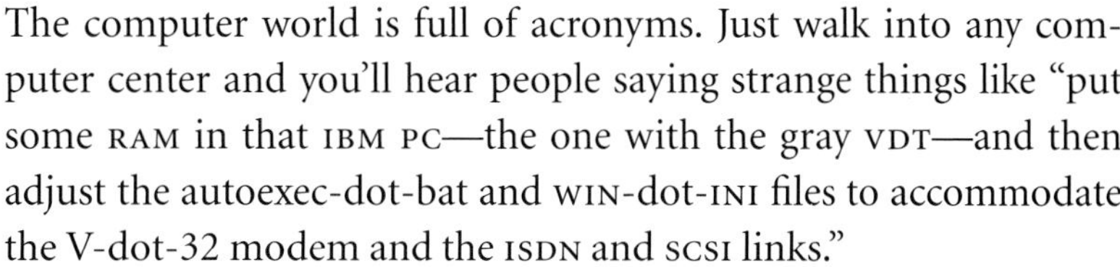

The computer world is full of acronyms. Just walk into any computer center and you'll hear people saying strange things like "put some RAM in that IBM PC—the one with the gray VDT—and then adjust the autoexec-dot-bat and WIN-dot-INI files to accommodate the V-dot-32 modem and the ISDN and SCSI links."

Perhaps in the future you'll be able to take courses in that language instead of French or Spanish in school. But for the time being, it's like a summer abroad: You've just got to pick up what you can as you go along. In this chapter we explain some of those weird acronyms that have to do with graphic file formats.

Depending on the types of graphics you're working with (and what manipulations you want to make), you'll have different requirements for the file formats storing those graphics. If you're capturing scanned images for high-quality color correction and output, for instance, you'll use a different format than someone creating drawings in AutoCAD and printing on a 300-dpi laser printer.

Placing, Printing, and Editing

When we talk about file formats, we're referring to the manner in which the graphic image is stored on the disk. How it's stored determines what programs can open ("read") the graphic, and what you can do with it once it's opened. If you say a program can read a given format, you may mean one of two things. It may be able to place the image on a page and print it (and perhaps apply some overall tonal adjustments), or it may be able to open the file and actually edit it.

Some file formats are industry standards, and others are proprietary: only the creating application can read and write the format. No other program that we know of, for instance, can read or write FreeHand's native file format. Proprietary formats have the advantage of being specialized for given programs, so they can store those programs' information completely and efficiently. Their disadvantage is that other programs can't read the information (for placing, printing, *or* editing) without translation.

EPS and TIFF are the two primary formats of interest to desktop publishers. For the sake of completeness, we also cover some formats that aren't used as often (but which you may end up having to work with), including PICT, CGM, MacPaint, and Windows Metafile.

Objects and Bitmaps

The first and most important distinction to understand with file formats is the difference between object-oriented and bitmapped graphics. Let's take a look at these first.

Bitmapped graphics. Bitmapped images are just big grids of dots (we call them *samples*). The samples can be different colors, and the number of dots can vary (as can their density, or resolution), but aside from that, a bitmap is just a map of bits of information. No matter what the picture is—whether it's a bar chart or a photograph—it's described using lots of dots. So a bitmapped file format needs to contain information on dimensions, resolution, bit depth, color model, and compression. (We cover the first four in Chapter 16, *Scanned Images*; compression is covered in Chapter 22, *Getting Small.*)

Bitmaps come from three primary sources: scanners; painting and image-processing programs (such as MacPaint, Windows Paint,

Photoshop, and so on); and screen-capture programs (Tiffany, Exposure Pro, and a host of others). If you create a graphic with any of these tools, it's a bitmap.

OBJECT-ORIENTED GRAPHICS. Object-oriented files are typically more complex. Instead of describing a rectangle with a bunch of dots, object-oriented graphics just say, "draw a rectangle this big and put it at such-and-such place on the page." These images can include lines, boxes, circles, curves, polygons, and text blocks, and all those items can have a variety of attributes—line weight, type formatting, fill color, graduated fills, *et al.*, *ad nauseam.*

There is no universally editable object-oriented file format that is robust enough to describe all those different sorts of objects and attributes (the closest competitor is the Illustrator 1.1 format; see below). In other words, you can almost always place an object-oriented graphic on your page and print it, but it's rare that you can edit it.

Object-oriented graphics come from two primary sources; drawing programs (CorelDraw, FreeHand, Canvas), and computer-aided design (CAD) programs. You also might get object-oriented graphics out of a graphing program, or from a spreadsheet with a graphing module.

BITMAPS QUA OBJECTS. As it turns out, the distinction between bitmaps and objects is slightly fuzzy, because object-oriented files can include bitmaps as objects in their own right. If you're creating an illustration in CorelDraw, for instance (or most other object-oriented drawing programs), you can include a scanned image as one of the objects in the illustration. The bitmapped image actually acts like an object on the page, much like a rectangle or oval.

If you include a bitmap as an object in an illustration, you can almost never edit that bitmap. You can place it, rotate it, or twist it around, but you can't actually go into the image and change it. Some programs, however, such as Canvas and SuperPaint, have object-oriented *and* bitmapped layers, allowing you to jump back and forth between models.

BITMAP-ONLY OBJECT-ORIENTED FILES. An object-oriented file can also include a bitmap as its *only* object. In this situation, the file is essentially a bitmapped graphic file, and depending on the file format, you may be able to open the bitmap for editing in a painting or image-processing application. (Photoshop's EPS files are the best example of this.)

SCREENING INSTRUCTIONS

Half of this book is about halftoning, so we can't neglect to talk about screening with different file formats. Remember, halftoning isn't just for scanned images; it's necessary for anything gray—lines, boxes, type or whatever. So you may want to specify screening instructions (frequency, angle, spot shape) within a file, whether it's bitmapped or object-oriented. If that's the case, you need to save the file in EPS format—the only format that can include screening instructions. (For more on how these internally specified screen settings interact with other programs' settings, see Chapter 13, *Controlling Halftone Screens.*)

TONAL ADJUSTMENTS

Another important consideration in choosing file formats is tonal adjustment. If you're planning on placing a scanned image on a page, then using the tools in your page-layout software to adjust tonal levels, you can't use EPS. The tonal values in an EPS file are fixed, and can't be altered or tweaked unless the image is opened for editing by an image-processing program.

If you want to use a page-layout program's tonal controls on a scanned image, you need to save the image in a bitmapped file format—preferably TIFF. Of course, you're probably better off managing your tonal control in an image processing program rather than relying on the lesser controls in page-makeup software. See Chapter 20, *Tonal Correction*, for more on tonal adjustments.

Bitmapped File Formats

The following discussions cover the most widely used bitmapped file formats (and some that aren't so widely used).

TIFF

The Tagged Image File Format (TIFF, pronounced just like it reads) is the most widely used, industry-standard bitmapped file format. A TIFF bitmap can be of any dimensions and any resolution (at least we haven't heard of any limits). It can have bit depths of one, four, eight, 24, or 32 bits per sample (four-and eight-bit TIFFs can be either color or grayscale), using grayscale, RGB (indexed or not), or CMYK color models, and can be saved in compressed and uncompressed formats. Almost every program that works with bitmaps can handle TIFF files—either placing, printing, correcting, or editing the bitmap.

TIFF sounds like the ideal bitmapped file format, but in fact the picture is not totally rosy. As it turns out, there are several different "flavors" of TIFF. It's such a flexible format that TIFFs written by some programs are incomprehensible to some others. This isn't as much a problem today as it was in past years when TIFFs varied widely, but there are still times when you have to open a TIFF in an intermediary program, then save it from that program in order to read it in another program.

IBM versus Mac. For some reason, the IBM and the Mac have different versions of TIFF. It has something to do with the file's byte order and the processing methods of Motorola (Macintosh) versus Intel (IBM) chips. For whatever reason, you sometimes need to convert TIFFs when you move them between platforms. Happily, programs like Photoshop (on both Mac and Windows), HiJaak, and DeBabelizer can read *and* write both Mac and IBM TIFFs.

Color Models. The latest TIFF specification (version 6) allows for CMYK color specs in addition to RGB. This means you can separate an image into the four process colors, and save that preseparated image in a TIFF file. When you place that file in a page-layout program or the like, no further separation is required. The program can simply pull the cyan channel when it's printing the cyan plate, the magenta channel when it's printing the magenta plate, and so on.

There is facility in TIFF files to use indexed color, but using indexed color is a prime cause of compatibility problems, in our experience. If we need to use an indexed-color TIFF, we immediately convert it to straight RGB or CMYK and damn the file size.

COMPRESSION. You can save TIFFs in uncompressed and compressed formats (using the LZW compression method, for those of you who understand that). If you want to use some other compression method, you'll need to use another format, such as EPS (for more on compression, see Chapter 22, *Getting Small*).

SCREENING INSTRUCTIONS (NOT!). There is no facility for including screening instructions in a TIFF file. Screening is controlled by the program from which the TIFF is printed. If you want to save a bitmap with screening instructions, use EPS.

RASTER IMAGE FILE FORMAT (RIFF)

RIFF is really hardly worth mentioning, except that you might run into it one day. RIFF is the preferred format of ImageStudio (for those of you who remember ImageStudio) and ColorStudio. If you run into a RIFF file, you need to open it in one of those programs (or a translation program like DeBabelizer) and save it as TIFF.

PCX

Whereas TIFF is an industry standard, the PCX format is controlled by ZSoft Corporation, the developers of Publisher's Paintbrush. It's a granddaddy of bitmapped formats, predating Windows 1.0 when it hit the streets as part of PC Paintbrush. The current version of PCX supports adjustable dimensions and resolutions, and 24-bit color, but only a 256-color palette (indexed to 24-bit color), up from earlier four- and 16-color versions.

Since a variety of palette-color techniques have been applied to PCX files over the ages, files from earlier programs can have serious color-mismatch problems. But if you're satisfied with the results of working with the PCX images you have, then go for it. We typically recommend using TIFF files instead of PCX whenever possible.

MACPAINT

The MacPaint format is ultimately the most basic of all graphic formats on the Macintosh (a couple of PC programs can open or place MacPaint files, but otherwise it's a nonissue on the PC). Paint files (more rarely called PNTG, or "pee-en-tee-gee," files) are black and white (one bit per sample), 72 samples per inch, eight-by-10 inches (576 by 720 dots). That's it. No more and no less.The MacPaint format is useful for capturing and placing black-and-white Mac screen shots (especially since it's so compatible with every Mac program), but otherwise we don't find much use for it.

WINDOWS BITMAP (BMP)

Windows Bitmap (typically called BMP, pronounced by saying the letters) is the bitmap format native to Windows Paint. It is rarely encountered outside of Windows and OS/2 Presentation Manager. You can store a one-, four-, eight-, or 24-bit image of various dimensions and resolutions. We still prefer TIFF, given its strong support by desktop-publishing applications and compatibility across different computer systems.

GRAPHICS INTERCHANGE FORMAT

The Graphics Interchange Format (commonly known as GIF, pronounced "jiff"), is the "house-brand" image file format of the CompuServe online information service. There are loads of free images available on CompuServe in GIF format—everything from satellite images to soft porn.

Most image-editing programs can now read and write GIF image files, but you'll probably have to convert them if you want to place them in page-layout or illustration programs. GIF only supports 256 indexed colors.

Object-Oriented File Formats

As we mentioned above, object-oriented file formats need to store a much wider variety of information than bitmapped formats. Given that variety, it's not surprising that there isn't a good industry standard object-oriented file format that can be edited by a wide variety of programs and printed to any printer. The closest thing is Illustrator 1.1-format files.

Bear in mind that you can use some of these object-oriented file formats as bitmap-only formats. This means that a painting or image-processing application can open the file for editing.

Encapsulated PostScript (EPS)

Encapsulated PostScript is absolutely the best format for storing graphics of any kind. That's a big statement, but it's true—as long as you have a PostScript printer, and as long as you don't want to edit the graphics—even to adjust their tonal values. With some exceptions, you can't print EPS files on non-PostScript printers, and you can't open them for editing. We discuss the exceptions below.

If you're like lots of people, you're wondering what encapsulated PostScript is, and why it has such a funny name. Let's start by talking about what PostScript is.

PostScript is a page-description language. It's like Basic or C or any other programming language, except that it's optimized for putting type and graphics on paper. When you're working with a PostScript printer and you tell your word processor (or any other program) to print a page, the computer writes a program in PostScript that describes the page, and sends that program to the printer. The printer, which actually has a full-blown computer and PostScript interpreter inside it, interprets that program and draws the picture on the page.

If you don't have a PostScript printer, you can't print an EPS graphic. The one exception is if you have a software PostScript interpreter on your computer. In that case, that program (such as Freedom of the Press) interprets the PostScript and turns it into a bunch of dots that the non-PostScript printer understands.

So at its lowest level, a PostScript file is just a bunch of text commands that a PostScript interpreter understands. EPS is a refined form of PostScript, and it comes in several varieties, including with and without preview, Illustrator 1.1 format, and bitmap-only EPS.

BINARY VERSUS ASCII. In all the varieties of EPS discussed below, there is one important option (which some programs let you set)—binary or ASCII encoding. This only applies to bitmapped images within EPS files—not to objects. The difference is simple: binary-encoded bitmaps are half the size of ASCII-encoded bitmaps. ASCII encoding is sometimes useful when you're transferring files over networks, or printing via serial cables.

EPS WITHOUT PREVIEW IMAGE. In their simplest form, EPS files are just made up of PostScript code that conforms to a few document structuring conventions defined by Adobe (creator of PostScript)—certain commands can't be used, and there are some comments in the file specifying the file name, creator, and so on.

If you get one of these files and place it on a page, you won't see the graphic; you'll just see a gray box, and probably some of the comments. For example, if you transfer an EPS image from a mainframe or workstation to your PC and then place it on your PageMaker page, you'll only see a gray box where the picture should appear. When you print to a PostScript printer, however, the graphic will appear in all its glory.

EPS images don't need previews on computer platforms based on display PostScript, such as Nextstep and some Silicon Graphics workstations, because their Display PostScript actually interprets the PostScript and displays it on the fly.

EPS WITH PREVIEW IMAGE. The most common form of EPS has a low-resolution screen-preview image built in. We call these viewable EPS files. When you place one of these files on a page, the computer shows you the low-res image, and when you print it, the computer uses the PostScript code. (If you don't have a PostScript printer, what prints out is the low-res preview—not terribly useful except for simple proofing.)

Unfortunately, viewable EPS files on IBMs and Macs use different screen preview formats (on the IBM it's either a Windows Metafile preview or a TIFF; on the Mac it's a PICT). So when you move an EPS from the Mac to the IBM or vice-versa and place it on a page, you often just get a gray box for a preview (some programs on the Macintosh, such as PageMaker and QuarkXPress, can now preview PC EPS files). Because the program uses the PostScript to print rather than the preview image, it should print properly.

Illustrator 1.1 format. Adobe Illustrator is a remarkable program, and one of the remarkable things about it is that it actually saves its images in PostScript rather than a proprietary format. However, it only uses a subset of PostScript commands. For better or for worse, the subset that Adobe Illustrator version 1.1 used has become the most widely used editable PostScript format around.

It's available with or without screen preview, and it has two advantages. First, it's quite robust—it can describe most of the object attributes necessary for sophisticated graphics. Second, many programs on both the Macintosh and the PC can open and edit the Illustrator 1.1 format (that includes Aldus FreeHand, CorelDraw, Arts & Letters, and of course, Adobe Illustrator).

It's a good format for passing around graphics that might need to be edited. A lot of clip art comes in this format. Unfortunately, Illustrator 1.1 EPS files can't include bitmaps.

Desktop Color Separation (DCS). The DCS specification was developed by Quark, Inc., for use in process-color work. A DCS image is an EPS format that consists of five parts: a low-resolution screen preview, plus cyan, magenta, yellow, and black layers.

DCS version 1.0 is made up of five files—the low-resolution version that you place on a page, which contains pointers to four data files. These are sometimes called five-file DCS or EPS-5 files. DCS version 2.0 format can contain all the information in a single file. Plus, DCS 2.0 files can contain more than just four process colors; you can have a number of spot colors included, as well.

PHOTOSHOP BITMAP-ONLY EPS. As we said earlier, in general, any bitmaps saved within an EPS file cannot be edited; you can only place them, rotate them, skew them, and print them. Typically, when you want to be able to edit a bitmap, you should use the TIFF format. The exception is the bitmap-only EPS files created by Photoshop. This program can open its own EPS images. Although some other programs are beginning to edit this bitmap-only EPS format, the format hasn't achieved the standardization that Illustrator 1.1 files have.

PICT

The PICT format (pronounced just like it looks) is the Mac-standard object-oriented file format. A PICT graphic can contain a bitmap as one of the objects in the file, or as its only object ("bitmap-only-PICTs"). Bitmap-only PICTs can be any size, resolution, and bit depth (the original PICT format only supported black and white, but that's no longer the case).

PICT is a respectable format for saving bitmaps that are in the editing process (especially since they can use QuickTime's compression options, making them smaller than some other formats). However, you should plan on converting to TIFF or EPS before placing the image in a page-layout package; PageMaker and QuarkXPress, at least, are limited in their ability to manage PICTs effectively.

Object-oriented PICTs have real reliability problems, especially when you're attempting high-quality output. Line widths and character spacing are especially subject to random changes—often resulting in real garbage on output. And gray fills are often represented by weird bitmap patterns. Nonetheless, PICT is the primary format when you're printing to non-PostScript devices.

COMPUTER GRAPHICS METAFILE (CGM)

Some years ago some industry/government group got together and tried to define a standard object-oriented file format. The result is CGM, and (as you might expect) it's not very standard, or very good. There's no facility for decent type or percentage tint specification, and CGM images can't include bitmaps at all. And since it's a extensible format, compatibility is a big problem with CGM files.

WINDOWS METAFILE (WMF)

The Windows Metafile—the standard object-oriented file under Windows—is a big improvement over CGM (from which it's descended). It handles type specification pretty well, supports percentage tints, and can even include bitmap graphics as objects. Unless you're printing to a non-PostScript printer, however, we strongly recommend the much more reliable EPS format.

MACINTOSH FILE TYPES AND DOS EXTENSIONS

One last thing you should know about file formats is that Macintoshes and PCs "see" files differently. On the Macintosh, every file has several attributes attached to it, including file type and creator. These are mysterious four-letter labels that tell the Macintosh what sort of file it is and what program generated it. For example, when you double-click on a file, the Mac looks at the file's creator to see what application to start up.

If you move an image from some other platform to the Macintosh, you may not be able to open or place it, because it has a file type of TEXT or DATA or some such, when the opening application is looking for EPSF or TIFF or the like. You can view and change a file's type using PrairieSoft's DiskTop, Apple's free ResEdit, and various other commercial, public-domain, and shareware programs.

In the PC world everything is different. There are no file types, no file creators . . . there are only file names. PC files all have names that are eight-dot-three. That means that the name can be no longer than eight letters, followed by a period, and ending with a three-letter extension. This extension provides all the information (and it ain't much) about the file's type and creator. Graphic files on the PC have extensions such as .TIF, .EPS, and .BMP.

Working with Files

Without a solid grounding in what kind of graphic files you're working with, it's almost impossible to be efficient in desktop publishing—you're forever banging into barriers. Why can't I import this image into my illustration program? Why can't I edit this photograph in this program? Why can't I move this file format over to the Macintosh? Why can't I use tonal correction on this file? And so on.

However, now that you know something about bitmapped versus object-oriented images, and TIFFs versus EPS files, you'll be able to really start to fly. Our next few chapters go into more detail in the area of bitmapped images, particularly scanned images.

18 Image Resolution

Scanning, Scaling, and Resampling

We come, at last, to perhaps the least understood topic in electronic publishing: resolution. The word resolution comes up in conversation—directly or indirectly—so often that you'd think it was a synonym for money. In some ways it is: if you don't understand resolution in all its different forms, you can't be efficient in your work; and if you aren't efficient in your work, you can't make money in this business.

The word resolution is used in so many different contexts that people quickly get confused about what it means. There's the resolution of an image, the resolution of an imagesetter, the resolution of a scanner, and so on. In this chapter we want to solidify your understanding of scanned image resolution in all its incarnations. Just remember what we keep saying: it's not as hard as it sometimes seems.

They're All Just Dots

Whenever someone talks about resolution, they're talking about a bunch of dots that are assigned or mapped to a given space (typically measured in inches or centimeters). It's as simple as that. The complicated part is getting it clear in your head whose dots are whose, and what those dots are doing.

Although most people talk about resolution using the phrase "dots per inch" (dpi), the only thing we specify in dpi is printer resolution. For scanned images or bitmapped images, we talk about samples per inch (spi); for display resolution, it's pixels per inch (ppi). And for halftone screens, of course, we talk about lines per inch (lpi). However, when you talk to other people, they might still use "dpi." Don't get confused; just correct them.

Also, many service bureaus are now using the term "res" to refer to the resolution of a scanned image. Res is another word for samples per millimeter. Scanner operators often scan at "res 12," meaning 12 samples per millimeter or 304 spi (25.4 mm to the inch).

Scanning Versus Image Resolution

If there's one distinction we'd like you to leave this chapter with, it's the difference between *scanning resolution* and *image resolution*. They're not always the same. Just because a scanner scans an image at 200 spi doesn't meant that the resolution of the image that finally gets printed will be 200 spi. The resolution can be affected by scaling and resampling.

Scaling

First and foremost, when you place a scanned image on a page and scale it, you change the image resolution (see Figure 18-1). Let's say you have a two-by-two-inch, 100-spi image. Reduce it to 50 percent, and you have a one-by-one-inch, 200-spi image (same number of samples packed into half the space).

If you specify a scaling percentage in the Print dialog box, you're further affecting resolution. If you enlarge to 200 percent, for instance, you're reducing the resolution by a factor of two.

FIGURE 18-1
Scaling and resolution

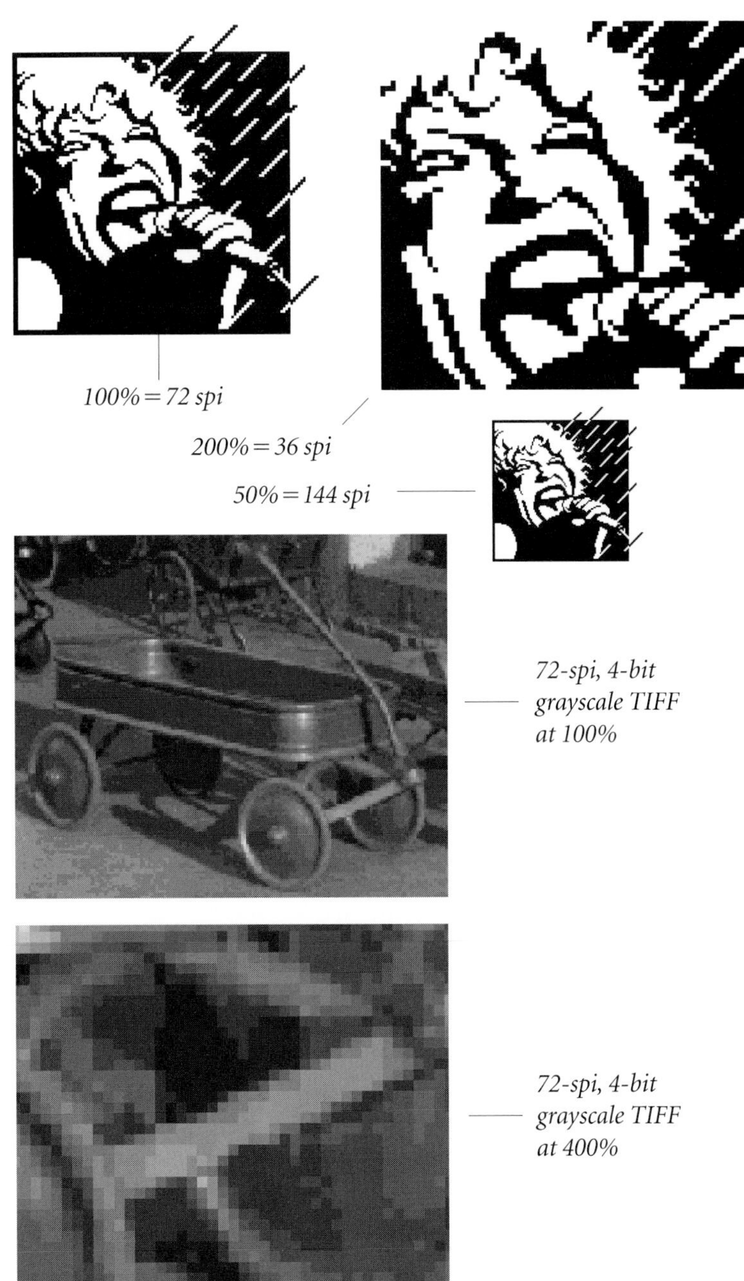

RESAMPLING

You can resize a bitmap without affecting resolution, however. Or you can change a bitmap's resolution without changing it's size. This process is called *resampling,* and it's accomplished using an image-manipulation program.

The best way to explain this is to show you Photoshop's Image Size dialog box (Figure 18-2). The top part of the dialog box shows you the size of the image before resampling, and the bottom shows the changed settings. Other programs offer similar controls.

Image Size...
Current Size: 198K
Width: 1.417 inches
Height: 1.59 inches
Resolution: 300 pixels/inch
New Size: 88K
Width: 1.417 (inches)
Height: 1.59 (inches)
Resolution: 200 (pixels/inch)
Constrain: ☒ Proportions ☐ File Size
OK
Cancel
Auto...

FIGURE 18-2 Photoshop's Image Size dialog box

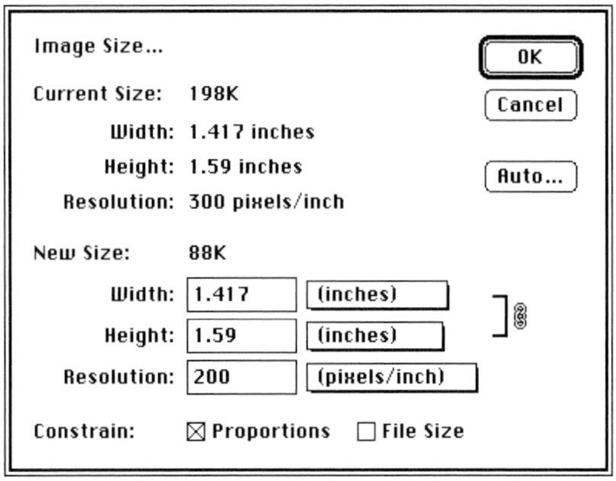

When you reduce the resolution of a bitmap, it's called *downsampling.* When you increase the resolution, it's called *interpolation.* In either case, there are smart and not-so-smart methods.

Interpolation is the process of generating a sample point in a bitmap where there wasn't one before. Downsampling is similar, but it removes sample points throughout the image. Most resampling processes are very basic. A scanning or image manipulation program might add a sample point, and to figure out what color it should be, just look at the sample next to it ("nearest neighbor"). Or, if the software is slightly smarter, it'll look at the samples on either side and average between the two ("bilinear"). Sometimes this process gives you a good image, and sometimes it doesn't. The best kind of interpolation is bicubic, which looks at all the surrounding samples and applies their average value to the new sample point. This takes the longest, but yields the best quality. Photoshop offers all three methods.

THE LIMITS OF INTERPOLATION

Unlike downsampling, which simply removes existing information (using more or less intelligent methods), interpolation is more problematic, because it's attempting to add information that doesn't exist in the original image (see Figure 18-3). While it can help reduce the aliasing (a.k.a *the jaggies*) in some situations, it can't add detail to an image if the detail isn't already there.

FIGURE 18-3 Interpolation

Firm Resolutions

With any luck, now that we've laid out the essentials of resolution, next time someone comes shaking their scan at you, you'll be able to fix their problem and send them on their way. Plus, to really get you on your way, in the next chapter we get into some specific advice on what resolution settings to use for different types of images.

19 Choosing Resolution

When Enough is Enough

Remember the good ol' days when cars had a choke knob? In the cool early morning you could help your car start by pulling the knob out a little and then perhaps pumping the gas pedal a bit to prime the engine. If you did too little, your car wouldn't start properly. If you pulled the knob out too far and pumped too hard, your engine would flood with gasoline and you'd sit there frustrated in the fumes. Unfortunately, cars not being digital, figuring out just how much to pull and pump was more an art than a science.

Enter scanning, stage left. If scanning is like an old car, then scanning resolution is like the choke knob. How do you know what resolution you should scan at? The highest resolution possible, right? No! You'll too often flood the engine (your computer) with too much information. The image file size gets too big, it takes too long to process or print (if it prints at all), and everything gets bogged down. Okay, so how about scanning at a really low resolution? Images look pixelated and computeresque—fine if that's your design goal, but problematic for most work.

Fortunately, although there is an art to scanning, much of your work can be figured out by science. There are simple formulas for deciding how much resolution you need for different types of images.

Remember: the resolution we're talking about here is *image resolution*—after an image has been placed on a page and scaled to fit (to understand the difference between image resolution and scanning resolution, see Chapter 18, *Image Resolution*).

Line Art

Back in Chapter 16, *Scanned Images,* we talked about bilevel, or line-art, images. These are always black and white; each sample point in the image is described with one bit of information. Bilevel scans are great for reproducing straightforward line drawings, and even images as complex as the finely-etched pictures in those old-fashioned Dover clip-art books.

To achieve line-art quality approaching that of photographic reproduction, and to avoid the jaggy look (Figure 19-1), you need two things: very high resolution, and sharpening. We discuss sharpening in Chapter 21, *A Sharper Image* (but we will say here that sharpening improves line-art scans *dramatically*). We'll cover the details of line-art resolution here.

200 spi

400 spi

800 spi

1,200 spi

Figure 19-1 Jaggy edges in line art

When printing line-art scans, you never need image resolution higher than your output resolution (a 300-spi image looks about the same as a 600-spi image if they're both printed on a 300-dpi printer), but if you're printing on an imagesetter, plan on using an image resolution of *at least* 800 spi—preferably 1,000 spi or more. Yes, you heard us right. Line art with an image resolution of less than 800 spi shows jaggies, even to casual observers at normal viewing distances. Note that if you're printing to a toner-based laser printer or onto porous paper such as newsprint, you can often get away with lower resolution scans (perhaps 400–600 spi).

But how do you create an 800-spi image if you've only got a 300-spi scanner? There are two answers, which you can use independently or in combination.

LARGE ORIGINALS. The simple solution is to use large original art. Scan at your scanner's maximum resolution, then scale the image down on pages, increasing the resolution (or, if your scanning software lets you, scale the image down and raise the resolution in one fell swoop at scan time). If your original art isn't big enough, you can enlarge it photographically (even using a high-quality photocopier), with minimal loss of quality.

COOL GRAYSCALE-TO-LINE-ART WORKAROUND. Here's a technique that you'd never think would work, but it does. This method uses the extra information captured in a grayscale scan, and converts that information into higher-resolution line art. You can actually increase your line-art resolution by a factor of two over your scanner's true optical resolution using this technique. Here's the step-by-step (thanks to Rob Cook, the genius behind Ofoto, for explaining this technique).

1. Scan the image in grayscale mode, at the maximum resolution your scanner provides (we'll assume 300 spi for this example).
2. Resample the image (interpolating) up to 600 spi (you may be able to do this with your scanning software's controls, but consider doing it in Photoshop instead; scanning software's interpolation routines vary in quality).
3. Adjust the threshold level (the point where gray pixels jump to white or black) if necessary, making lines thinner and thicker in the process. You can do this by using the brightness control.
4. Sharpen the image (again, your scanning software may be able to do this as you're scanning; otherwise, use Photoshop's Sharpen or Sharpen More filters).
5. Convert the image to black-and-white (with no halftoning or dithering).

You've now got a 600-spi line-art scan, created with a 300-spi scanner! Note that some scanning software packages (such as Hewlett-Packard's DeskScan and Light Source's Ofoto) can perform this whole process automatically.

HIGH-RES LINE-ART OUTPUT

The big problem with these very high-resolution line-art scans is printing them. PostScript RIPs seem to be much more sensitive to the high resolution of these images than they are to overall image size. It's not uncommon to print 10-megabyte grayscale images (with resolutions of 200–300 spi) on imagesetters with no difficulty. When you get into 800- and 1,000-spi images, however, no matter their total size, imagesetters tend to bog down, and sometimes choke. The best advice here is to find an imagesetter with the fastest, most powerful, most up-to-date RIP available. And if you're producing a big job—especially with multiple scans per page—run some test pages first to see if you're going to run into problems.

On some imagesetters (notably Agfa's), a black-and-white bitmapped image prints much more quickly if it's at the same resolution as the imagesetter itself. The imagesetter doesn't have to rescale it, as it would even with a 600-spi image, so the bitmap just gets sent right through.

GRAYSCALE AND COLOR

As with line art, with grayscale and color ("deep") bitmaps, there's a relatively straightforward formula for deciding how much resolution you need: It should never be more than two times the output screen frequency (see Figure 19-2).

For example, if you're printing a halftone at 133 lpi, the image resolution shouldn't be more than 266 spi. Any higher resolution is just wasted information. Contrary to what you might think, it doesn't improve image quality at all (if you're looking for a solution to blurry-looking scans, increased resolution isn't it; sharpening is). And since increasing resolution increases file size geometrically (see Chapter 16, *Scanned Images*), it's well worth limiting resolution.

FIGURE 19-2
Grayscale images at 133 lpi

72 spi

133 spi

186 spi

266 spi

In fact, the two-times rule has become so common that people have forgotten (or never learned) that you can usually get away with lower resolutions. We typically recommend a multiplier of 1.5 or even 1.4, and for lower-quality images 1.2 often suffices. You can lose image detail at these lower resolutions, however, and run into two other quality problems—aliasing and mottling.

Aliasing. Hard diagonal edges in lower-resolution images often display aliasing, or jaggies. A sailing ship's rigging in front of a bright sky, for instance, can often look quite "stair-steppy" at lower resolutions. This effect is accentuated by sharpening. If hard diagonal edges or fine lines, like those found in a baby's hair, make up important elements in the image, you should consider using a higher resolution.

Mottling. Lower resolution scans can sometimes display mottling in areas of smooth gradation. In people's faces, it looks like a poor complexion. This is mainly a problem with scans from lower-quality scanners that have a lot of noise—samples in uniform gray areas that don't match their surroundings. Since sharpening works by accentuating the differences between adjacent light and dark pixels, it increases the mottled effect.

See Chapter 21, *A Sharper Image*, for more on addressing these problems.

AUTOMATIC DOWNSAMPLING IN PAGEMAKER AND QUARKXPRESS

Both PageMaker and QuarkXPress will automatically reduce the resolution of images while printing, downsampling them to two times screen frequency. In PageMaker, this is accomplished through the Optimized option in the Print dialog box. In QuarkXPress, it's automatic, undocumented, and—as of this writing—you can only turn it off with a separate XTension.

In either case, we don't know what resampling method the programs use, and would prefer to handle image resolution issues ourselves. The solution? Adjust the size and resolution of images in a program like Photoshop before importing them into a page layout program. If the resolution isn't higher than two times the screen frequency, you don't have to worry about it.

Resolution and Quality

We hope this chapter has convinced you to drive your computer like you'd drive your old beater car. It's hard to go wrong; just pay attention to the signs and don't try to throw too much information at the computer by flooding it with scans at too high a resolution.

In the next couple of chapters we want to look at some ways that you can enhance the quality of your scanned images: tonal correction and sharpening. These two topics are in many ways the keys to getting great-looking images out of your desktop computer.

20 Tonal Correction

Getting Great Grays

When you buy a new car, you make adjustments to it before driving it away. First you adjust the seat, then the mirrors, and you might even change the settings on the radio to the channels you listen to most. When scanning images, you almost never just roll with what the scanner gives you by default. You adjust the picture's tones, sharpen the image, and along the way you might even change parts of the picture to suit your needs.

In this chapter, we'll discuss one of the first things you want to adjust in your image: tone. You can think of tone as how the colors or gray levels throughout a picture relate to one another.

As we pointed out in Chapter 15, *Scanners*, many scanners have difficulty picking up detail in shadow areas. That means most of the desktop-created scans you see are too dark, with no detail at all in shadow areas. In fact, you almost never see a desktop-created scan that is too light. In this chapter we concentrate on dealing with that central problem—increasing shadow detail (and lightening the image overall) without washing all the highlight areas out to white, losing detail there.

Histograms

When Steve hears someone talking about histograms, he always thinks they're talking about allergy medicine. They're not. They're talking about an extremely useful tool for looking at how gray levels are spread out through an image. A histogram is simply a chart of how many samples in the bitmapped image are set to each gray level (see Figure 20-1). For example, a histogram can tell us that there are 10 sample points that are totally black, 34 sample points that have a gray level of 1 (out of 255 levels of gray), 40 sample points that have a gray level of 2, and so on.

By looking at the histogram of an image, you can quickly tell where the information in the image is concentrated—in shadow areas, in highlights, or in the midtones (see Figure 20-2).

Gray-Map Curves

In order to change the tonal balance of an image, you need to adjust its gray levels. Different programs let you do this is different ways, but typically the mechanism is through a *contrast* or *gray-map curve* (see Figure 20-3). It *maps* grays to different gray values.

Figure 20-1
Histogram

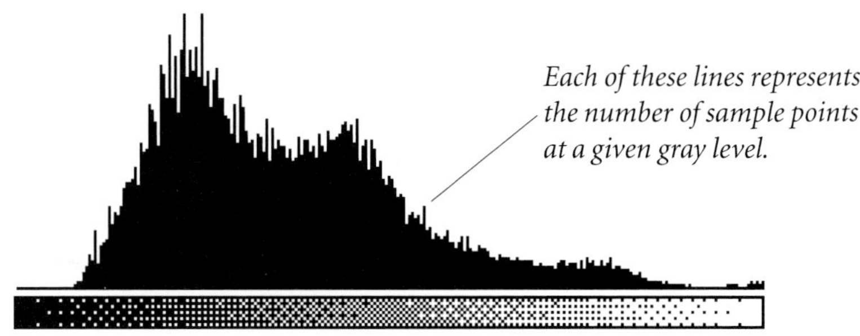

Figure 20-2
Various histograms

Original image

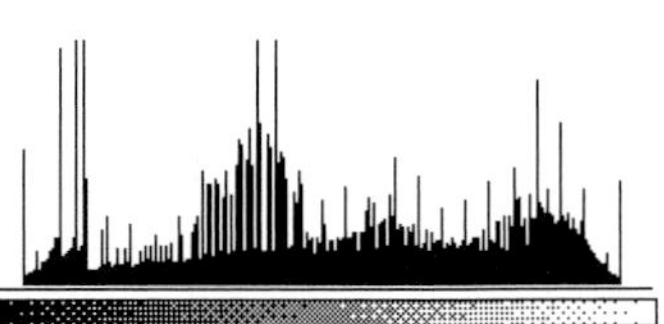

FIGURE 20–2
Histograms continued

Low contrast; tonal compression

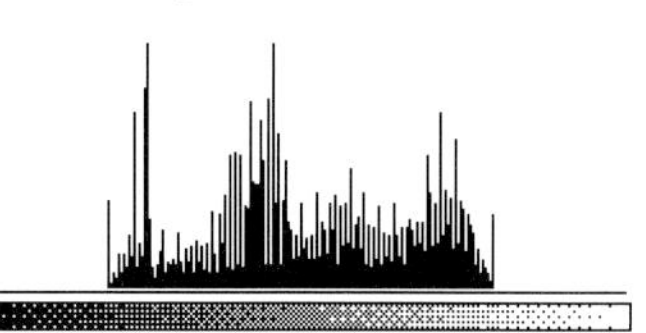

High contrast; expanded tonal range

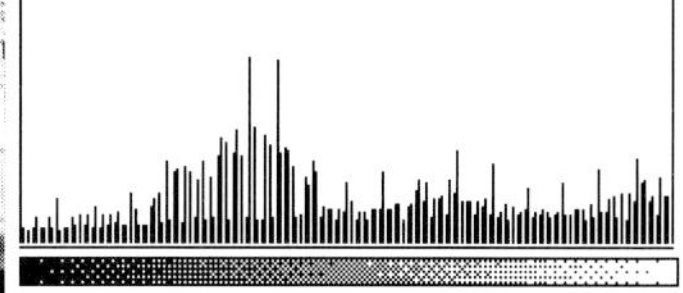

Darkened; histogram moves to the left

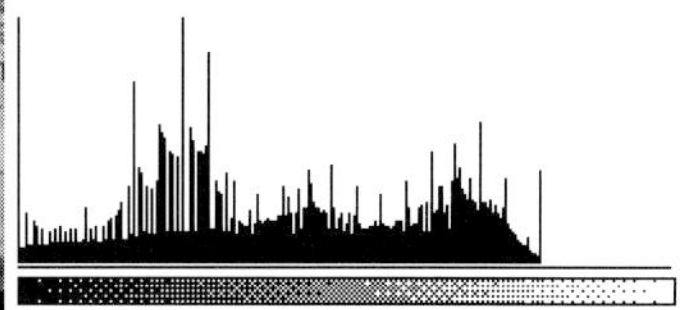

Lightened; histogram moves to the right

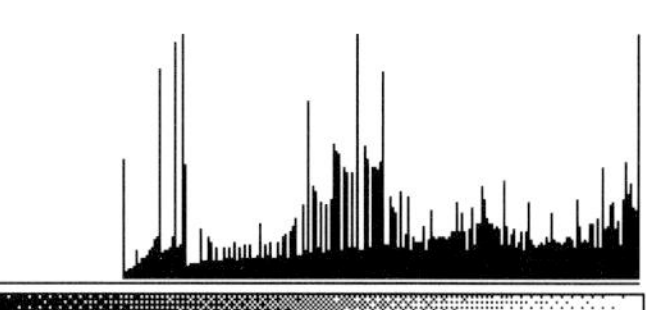

FIGURE 20-3
Linear gray maps

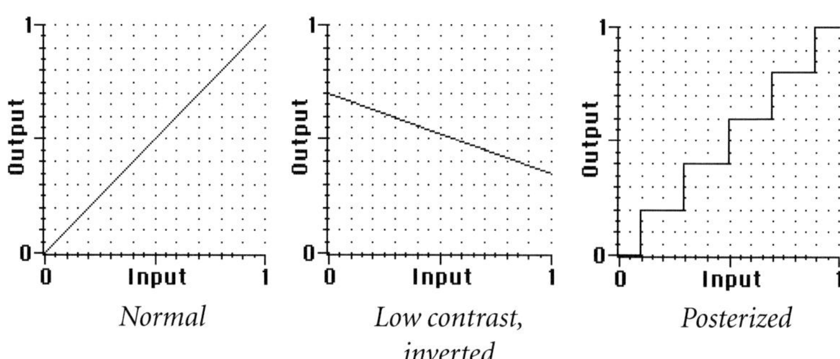

Perhaps the easiest way to understand a gray-map curve is to think about coffee or tea filters. When you pour water through a filter with coffee or tea in it, the water becomes flavored. You can use filters with images, too. And a gray-map curve is a type of filter that you can pour an image through.

By looking at the gray map, you can actually figure out what will happen to an image when you apply that filter to it. Typically, the horizontal axis of the chart describes the levels of gray—from white to black—of the input image (the image that you start out with). The vertical axis describes the levels of gray of the output image (what you'll end up with). A map with a straight 45-degree line in it like the one shown in the last figure is called a "normal" contrast curve. It doesn't change anything. After applying this filter, the pixels in the input image that are black stay black; the pixels that are 50-percent gray stay that tint, and so on.

Linear Correction

The most basic method of tonal correction is altering the slope of the line, while keeping it straight. Because the gray map is still a straight line, we call this *linear correction.* When you use run-of-the-mill brightness and contrast controls in a program, you are making linear corrections (even if you can't see the graph in your application, that's what's happening behind the scenes).

As you'll see in the next few paragraphs, linear correction throws away data in a not-terribly-intelligent manner. We consider linear correction—brightness and contrast controls—to be essentially worthless; but we'll explain how it works so you can agree with us.

You can effect linear correction in two ways: by moving the line up and down, and by changing the slope of the line.

BRIGHTNESS

The first linear correction we'll look at is adjusting brightness. If you move the line straight up, you lighten the image (see Figure 20-4; remember that some curves are just the opposite—it depends on how the application works). That's because all the sample points are made brighter. For example, let's look at a sample point that is 50-percent gray in our input image. When we pass it through this filter, that sample point gets converted into a lighter value—let's say 30-percent gray. Lowering the line has the opposite effect: it darkens the image. That same 50-percent sample point gets changed to a darker value when it is passed through the filter.

As you can see in the histograms in Figure 20-2, brightness controls just shift all the values in the image up. While brightening can make shadow detail easier to see, it doesn't increase the differentiation between subtly different values (which is what really increases detail in an image). And—this is the worst thing about linear brightness adjustment—all the information at the highlight end of the histogram simply disappears. Subtle tonal differences in highlights—the details—wash out to white.

CONTRAST

While changing the position of the line changes brightness, changing the slope of the line alters the image's contrast. If you make the slope of the line steeper, you increase the contrast of the image (see Figure 20-5). Gray levels that are nearly white get pushed to white, and those that are nearly black get pushed to black. And all the grays in between are spread apart, increasing the contrast between samples. If you make the slope of the line less steep than the 45-degree normal curve, you decrease the contrast of the image. This is also called tone compression, because you're compressing the whole range of 256 gray levels into a smaller number of gray levels.

FIGURE 20-4
Linear brightness adjustment

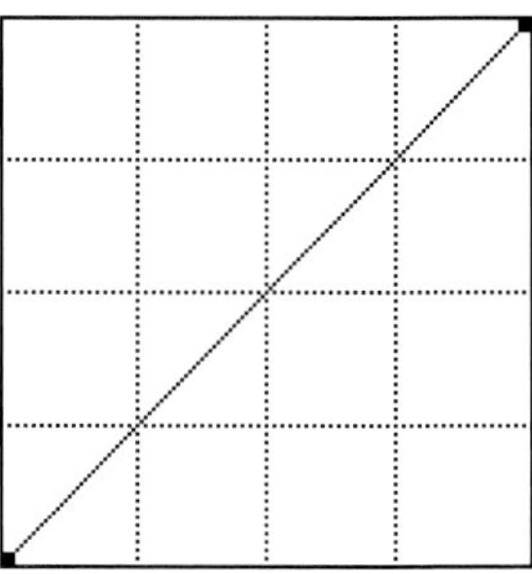

Before linear correction

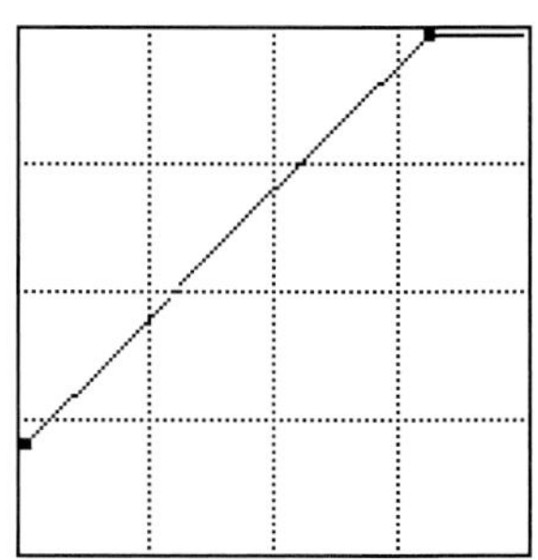

Line pushed up lightens the image

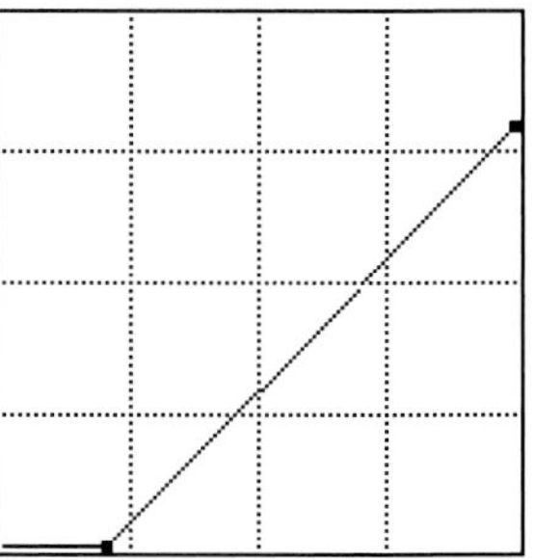

Line pushed down darkens the image

Linear contrast adjustment is just as problematic as linear brightness adjustment, but in a different way. Increasing contrast does increase the difference between adjacent gray levels—improving detail in the midtones—but it lops off detail in both highlights *and*

FIGURE 20-5 Linear contrast adjustment

Steeper slope, higher contrast

Shallower slope, less contrast

shadows. Decreasing contrast is almost never a very useful technique, unless you're seeking to limit the range of highlights and shadows for reproduction reasons (see Chapter 4, *Reproducing Halftones*).

NONLINEAR CORRECTION

To achieve the kind of tonal correction you really need—increasing shadow detail and brightening images without washing out highlight areas—you need to use nonlinear correction (sometimes called *gamma* correction).

Nonlinear correction is not a panacea, however. It loses data just as linear correction does, but in a much more sensible manner. When you use nonlinear correction to bring out shadow detail, you do lose

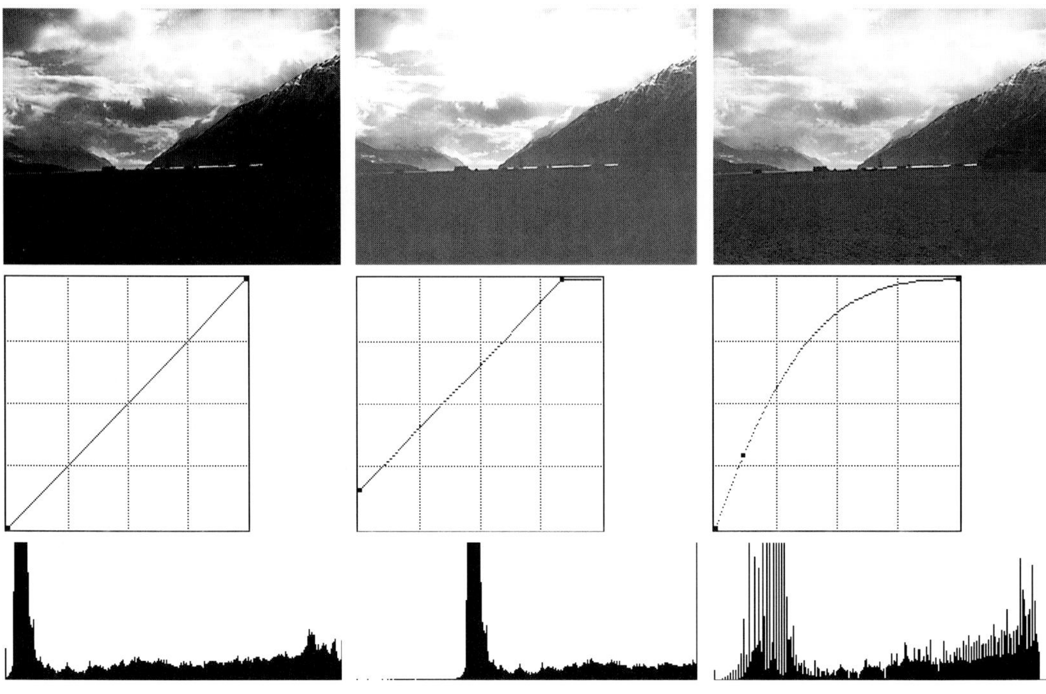

Uncorrected. As scanned, the image is very dark overall. There is no visible detail in the shadow areas—or well into the midtones, for that matter. (We chose this image because of its huge tonal range, with lots of subtle detail in both highlights and shadows.)

Linear brightness adjustment. This method simply pushes all the values up (the whole histogram is just shifted to the right), without increasing the differentiation between samples. All detail in the highlights is gone—pushed to white—but shadow detail is not enhanced much.

Nonlinear adjustment. The bars in the histogram are spread out, increasing the difference between samples. This increases visible detail in the shadows and midtones, with only minor loss of highlight detail.

FIGURE 20-6 Non-linear tonal correction

some detail in the highlights. Because the curve is flatter in the highlight area, samples that did have different values have the same values after correction. Since the biggest problem with desktop scans is bringing out shadow detail, however, the tradeoff is well worth it.

Another problem to watch for with nonlinear correction is posterization in deep shadow areas. If the angle of the curve in shadow areas gets too steep, you start seeing distinct jumps from one gray level to another—especially in areas of smooth tonal gradation.

METHODS OF NONLINEAR CORRECTION

Different programs offer different methods for nonlinear correction—from drawing a curve to moving sliders.

Drawing curves. A common, but not terribly useful, approach is the MacPaint-like pencil-and-graph tools in programs such as Digital Darkroom and QuarkXPress (see Figure 20-7). The lack of numerical readouts and inability to interact with a histogram make it difficult or impossible to predict or repeat results with any precision. It's also quite difficult to draw a decent curve with the little pencil (though the option offered in some programs to smooth your coarsely-drawn line can help solve this problem).

Figure 20-7
Pencil-like curve tools

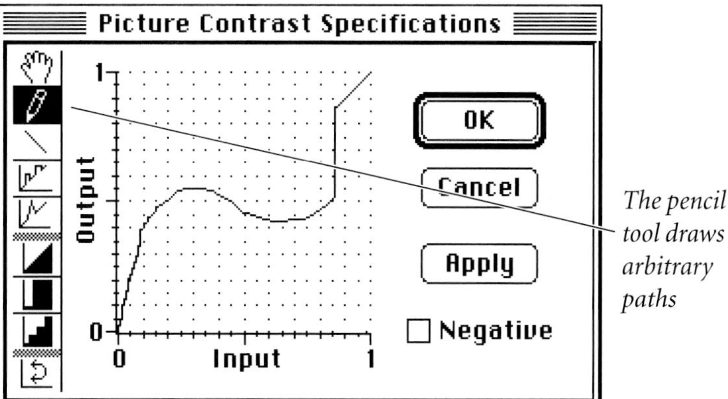

Adjusting curves' control points. A better method—one used by Adobe Photoshop, PageMaker, and QuarkXPress—is to let you move control points on the curve. It's more like working in a drawing program than a painting program. In the better-implemented of these techniques—Photoshop's, for instance—you can see the numeric values associated with control points as you move them (see Figure 20-8).

Adjusting input and output sliders. Adobe Photoshop and Aldus PhotoStyler (and others, too) take a different approach. You

FIGURE 20-8
Control-point curves

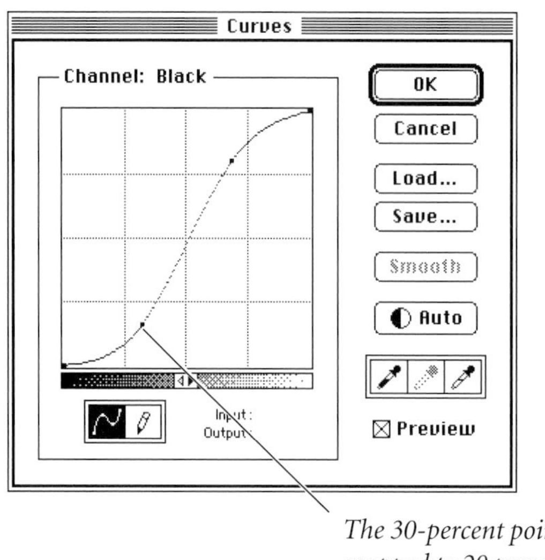

view a histogram of the image, and adjust sliders to remap gray values (see Figures 20-9 and 20-10). PhotoStyler even lets you see the gray-map curve at the same time. Whether it's Photoshop, PhotoStyler, or another program with a similar method, we like this technique the best. It's fast, easy, and does exactly what you need.

Here are the steps we typically go through to adjust grayscale and color scans in Photoshop (see Figure 20-10). PhotoStyler works

FIGURE 20-9
PhotoStyler's Gray/Color Correction dialog box

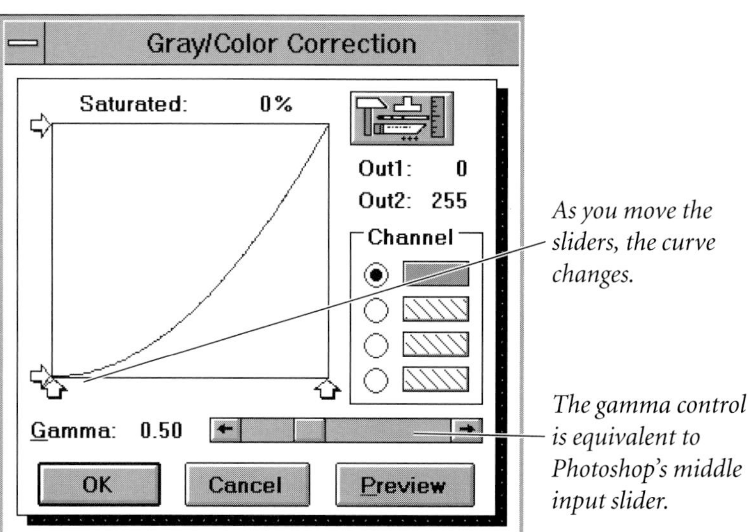

FIGURE 20-10
Using Photoshop's Levels dialog box

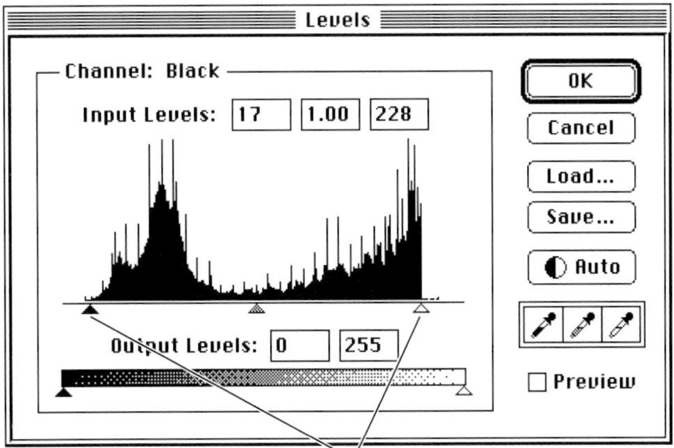

Move the left and right sliders to encompass the bulk of the information in the scan.

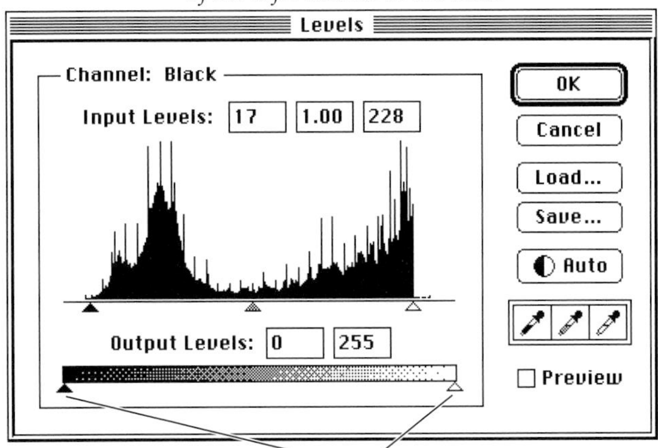

Move the output sliders to the values specified in Table 20-1 (or type in the values)

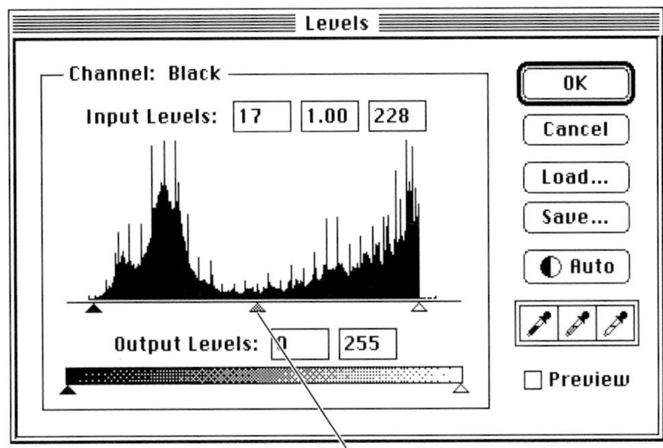

Move the center input slider to the left to brighten the image and bring out shadow detail.

similarly. Before you start working with Photoshop's Levels control, it's useful to make sure that Video LUT Animation is turned on in the General Preferences dialog box, and that the Preview checkbox in the Levels dialog box is turned off. That allows you to see the tonal corrections in "real time," as you make them.

1. Encompass the majority of the data on the histogram with the left and right input sliders. (See "Tone Clipping," below, for a useful technique for adjusting these sliders).

2. Adjust the output levels so that the resulting image doesn't include highlight or shadow dots that can't be reproduced on press (see Chapter 4, *Reproducing Halftones*). Table 20-1 shows typical ranges for three classes of printing paper. The third column does the arithmetic for you; it tells you what output levels you should specify in the Levels dialog box, though the principle applies to any scanning or image editing software. These values are just rules of thumb; talk to your printer to determine the best settings for your printing method.

TABLE 20-1
Output level settings

Printing Stock	*Percent Range*	*Output Levels*
Newsprint	12–88	30–225
Uncoated stock	10–90	25–230
Coated stock	5–95	12–243

3. Move the center input slider to the left until the image looks about right on screen. We usually bump this up to at least 1.5. Once it looks just about right, bump it up a little more. Scans look lighter on monitors (which are actually projecting light) than they do on paper (which relies on reflected light). You'll almost never make an image too light with this method; as we said above, almost every scanned image that comes off the desktop is too dark (see "Bit Depth and Dynamic Range" in Chapter 15, *Scanners*). Watch out for posterization in the deep shadow areas, however.

TONE CLIPPING

One of the most useful tools for tonal adjustment we've seen (and we've still only seen it in two programs—Photoshop and Cachet) is the tone clipping display (see Figure 20-11). This tool shows you which samples in an image are getting clipped to the tonal edges—going all black or all white.

Using the tone clipping display, you can make sensible judgements about what information is being lost, and how important it is. For example, your thinking might go like this: "Well, that shiny area over there is getting clipped to white, but it doesn't matter much because it's not central to the photo. I'll bring the slider in a bit more because I don't care if I lose detail there, and I can bring out more shadow detail by losing the information in that unimportant highlight area."

FIGURE 20-11
Photoshop's tone clipping display

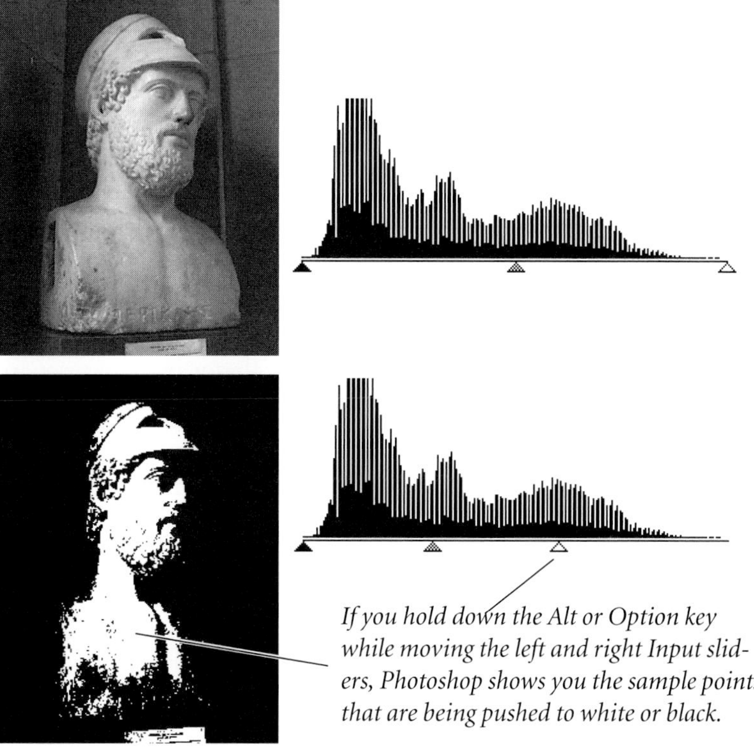

To use Photoshop's tone clipping display, hold down Option (Macintosh) or Alt (PC) while dragging the left and right input sliders in the Levels dialog box. In Cachet there's a Clip button you can press to see tone clipping.

EQUALIZE COMMANDS

Several programs offer commands that go under names like *equalize* or *balance.* Their purpose is to spread out the values in an image, so they are spread more uniformly across the range from white to black.

Most equalize commands, in our experience, are of very little use. On the other hand, there's at least one—Aldus PrePrint's Balance to Sample command—that does a remarkable job of adjusting an image for you automatically. Just select a neutral highlight area and choose Balance to Sample, and PrePrint does the rest. It's so good and so simple that our cohort Ole Kvern calls it the "Make Better" command.

CUMULATIVE CORRECTIONS

As we said earlier, making tonal corrections always involves some data loss in your picture. Sample points that were originally a range of different colors may become the same color after tonal correction; therefore, you lose detail in the image. With linear adjustment, you lose detail in highlights and shadows (they go all white or all black). With nonlinear adjustment, you lose detail mainly in the highlights, because the curve is flatter in that area. Samples that are different before correction are the same (or more similar) after.

Whatever correction method you use, remember that multiple corrections multiply the data loss. So if you use your scanning software's adjustment tools to correct an image, then use Photoshop's Levels dialog box to further adjust it, you lose data twice.

Some programs (Cachet and Live Picture, for instance) address this problem by "remembering" each of your corrections. They don't actually apply the corrections to the file when you make them; they just show their effects on screen. Then when you're finished correcting the image, they concatenate all the corrections into a single gray-

or color-mapping operation, and perform the actual correction only once. This is the best method of correction (one we'd really like to see implemented in Photoshop).

CLOSED-LOOP CALIBRATION

Several scanning programs (Hewlett Packard's DeskScan and Light Source's Ofoto among them) offer a method of non-linear tonal correction called *closed-loop calibration.* With this technique, you print a sample grid of gray or color bars from the program on your final output device. Then, when you scan that output, the application can build a special gray-map curve based on what it knows about the original image.

You can then apply that gray map to subsequent scans. When you do this, your final output should match the original scanned image. You can even extend the loop all the way through printing, so the whole process is calibrated, from scanning through ink on paper. In theory, at least, what comes off the printing press should exactly match the original photo.

This is a great approach—assuming you want the final output to match the original image. However, if you plan on using other tools to correct the image (if the original is too dark or too green, for instance), you face the multiple-correction data loss problem discussed above. There's some data lost when the image is corrected by the calibration routine, and some more is lost when you correct it manually. The loss may not be significant, but you should think twice about using closed-loop calibration if you plan on manually correcting the image anyway. (Ofoto concatenates the closed-loop calibration with any manual tonal corrections, however.)

KEEPING IN TONE

Almost every image that you scan in will need to be adjusted in a number of ways, including tonal correction. Without this, your scans will almost always be too dark and too muddy. Of course, as we've said repeatedly throughout this chapter, you want to keep as much image data as you can while you're correcting (an image's data is its *sine qua non*; when you lose the data, you've just got a pile of dirt left).

In the next chapter we're going to look at the second major image adjustment that you'll need to make in almost every scan: sharpening. Using these two corrections, you can create halftones that rival those from almost any high-end system.

21 A Sharper Image

Getting Rid of the Blurries

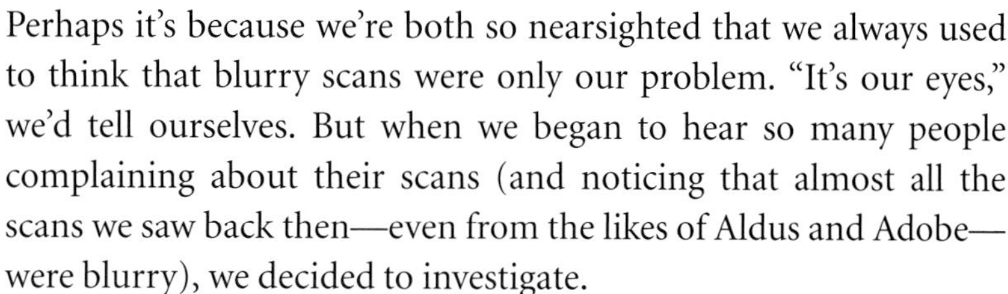

Perhaps it's because we're both so nearsighted that we always used to think that blurry scans were only our problem. "It's our eyes," we'd tell ourselves. But when we began to hear so many people complaining about their scans (and noticing that almost all the scans we saw back then—even from the likes of Aldus and Adobe—were blurry), we decided to investigate.

Resolution doesn't help. Steve spent years thinking that if he could just get high enough resolution, he'd be able to solve the problem of blurry scans. Well, he finally tested that theory, and it was plain wrong. Image resolution greater than two times screen frequency is simply wasted data; and you can often use much lower resolution images with excellent results (see Chapter 19, *Choosing Resolution*). Unfortunately, after five years of searching, we still don't really know why desktop-produced scans are so blurry.

However, we have learned how to make them as sharp and crisp as photographic halftones: software sharpening filters. This chapter is all about sharpening—the other essential manipulation (after nonlinear correction) for creating top-notch halftones of scanned images. Using software sharpening filters, you can often make an image even sharper than the original. And you can also oversharpen a picture to the point of degenerating the image. Let's see how it's done.

How Sharpening Works

Software sharpening filters work by accentuating the differences between adjacent light and dark sample points in an image. Since those areas where light and dark samples are next to each other are "edges" in an image, increasing the tonal difference between the samples increases the appearance of sharp edges (see Figure 21-1).

Figure 21-1
How sharpening works

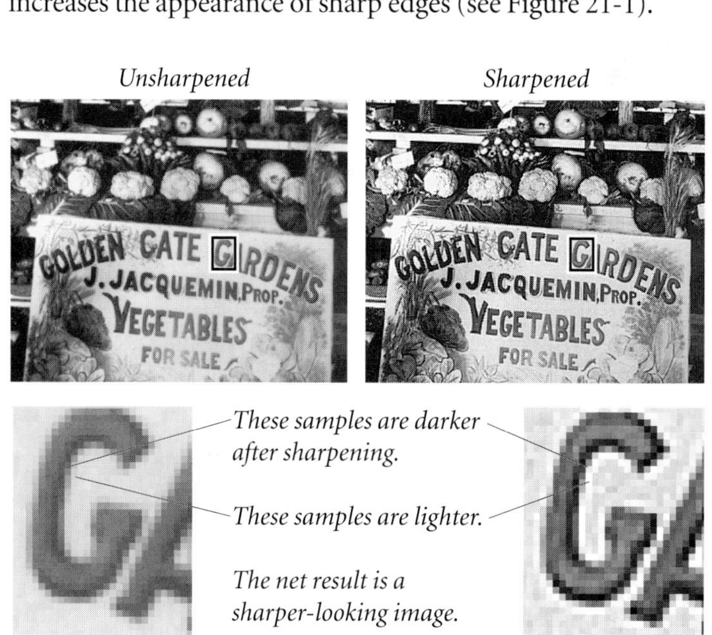

This straightforward approach to sharpening (the filters are usually called Sharpen or Sharpen More) works very well when you're sharpening grayscale images prior to converting them to line art (see "Cool grayscale-to-line-art workaround" in Chapter 19, *Choosing Resolution.*") For true grayscale and color work, however, the standard sharpening filters tend to accentuate details that you don't want to accentuate, like speckling in smoothly graduated areas (such as human faces or clouds) and scanning artifacts.

You can avoid these problems to some extent by using an edge sharpening filter. These filters don't accentuate small differences in adjacent tonal values—only large differences. So smoothly graduated areas don't get sharpened, but edges do. This is an improvement, but it still doesn't offer the controls you really need.

Unsharp Masking

The method of choice for sharpening scanned photographs is called (somewhat oxymoronically) *unsharp masking* (USM). With its roots in the darkroom and its genesis in high-end prepress systems, unsharp masking (in its best implementations, as in Photoshop), provides controls that let you adjust sharpening for particular situations.

Unsharp masking works by combining a slightly blurry (unsharp) version of an image with the original. This combination results in sharp details in high-contrast areas (the edges, where adjacent light and dark samples are markedly different), without accenting tonal shifts in low-contrast areas (areas of smooth gradation, where rapid tonal shifts would destroy the subtle transitions).

Since it increases the contrast in many areas of an image, unsharp masking also tends to increase the overall appearance of contrast.

How Unsharp Masking Works

How does combining a blurry version with the original make for a sharper image? We knew you'd ask that, so we prepared ourselves. Here's the skinny on how unsharp masking works.

Figure 21-2 shows a simple situation—a jump in tonal value from 40 to 60 percent. The accompanying graph depicts this value jump.

When you apply a blurring filter to the image (in our example we've used a Gaussian blur filter), you get what you see in Figure

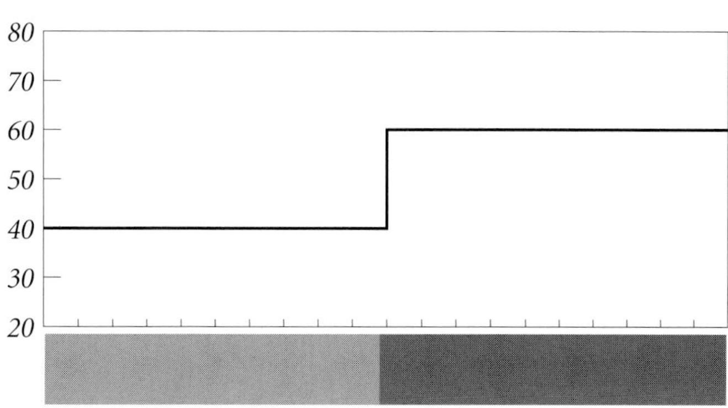

Figure 21-2
40 to 60 percent tonal jump

21-3. The values ramp up from 40 to 60 percent in a smooth transition. Note that this is an intentionally low-resolution bitmap we're working with, so it shows the effects clearly.

FIGURE 21-3
After gaussian blur

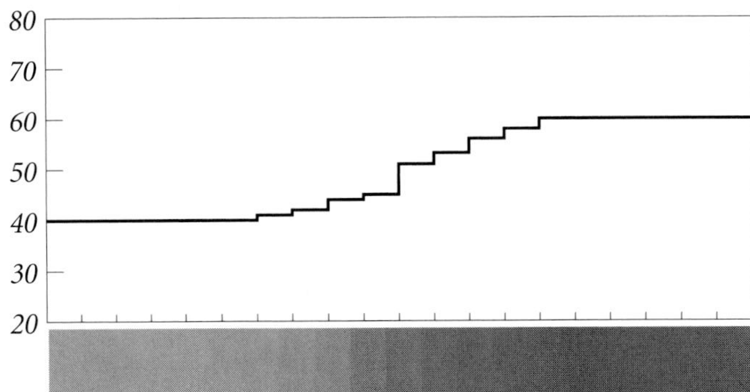

Finally, the filter does some clever if/then calculations: if the sample in the blurred image is darker than the same sample in the original, it makes that sample lighter by an amount equal to the difference. If the sample in the blurred version is lighter, it does the opposite. The result is an image like the one in Figure 21-4.

FIGURE 21-4
After unsharp masking

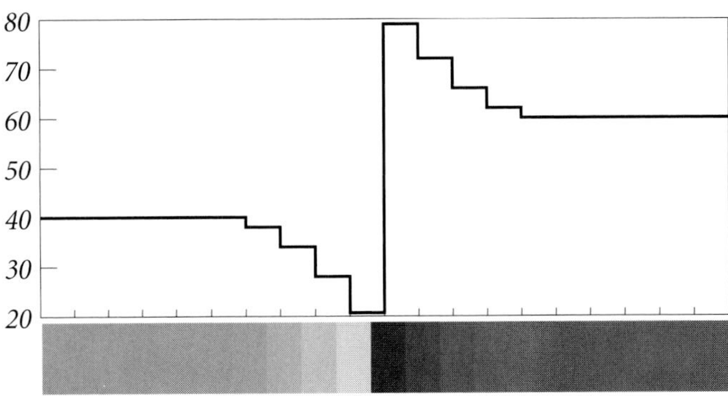

The key thing to notice in this figure is the "blips" on either side of the tonal shift. By accentuating the adjacent light and dark areas, unsharp masking gives the impression of sharpness. Figure 21-5 shows a photograph with intentionally exaggerated unsharp masking, to show the "halo" effect that unsharp masking creates.

FIGURE 21-5
The "halo" effect of unsharp masking

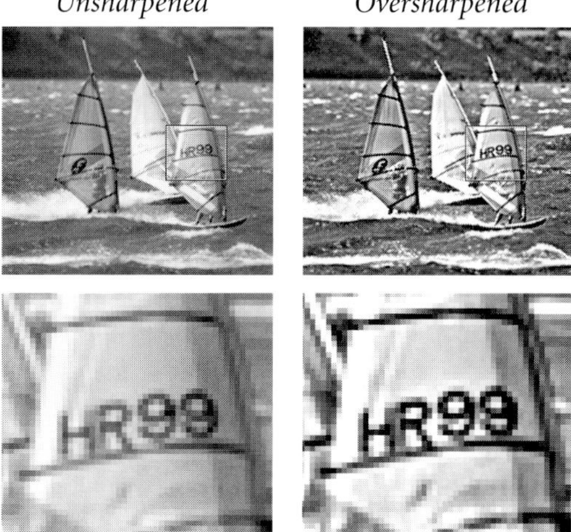

Note that the size of the halo is especially noticeable—not just the amount of tonal difference.

This halo effect is both the secret of sharpening, and its most potent Achilles' heel—depending on the size and intensity of the halo, and where in an image it appears. In the next section we'll discuss some of the problems that arise with sharpening; then we'll get into the details of Photoshop's Unsharp Mask filter, and discuss settings that optimize sharpening while avoiding sharpening problems.

PROBLEMS WITH UNSHARP MASKING

When used improperly or excessively, unsharp masking can make an image look artificial or bizarre, and it can accentuate problems in the original image or the scan.

HALOS. The blips that appear on either side of tonal shifts with unsharp masking can turn into a too-obvious halo, giving the image an artificial look. In excess, it can make an image look blurrier, as well. This is especially a problem when the halo is too wide, though it's also a problem if the blips are too light or too dark—if they're too different from their surroundings. Figure 21-6 shows an image that is sharpened excessively (though not as badly as Figure 21-5).

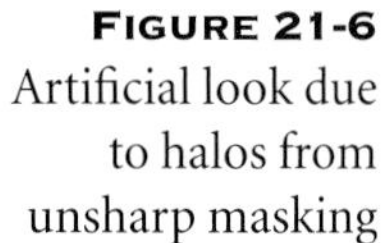

FIGURE 21-6 Artificial look due to halos from unsharp masking

ALIASING. Hard diagonal edges—like a dark building against a light sky—often display aliasing (the *jaggies*) after unsharp masking has been applied (see Figure 21-7). Aliasing is more of a problem with lower resolution scans—those that are less than two times the screen frequency (the lower the resolution, the more potential aliasing).

SPECKLING AND MOTTLING. Scanned images often contain a few random sample points that don't match their surroundings. This may be due to a noisy scan, irregularities in the original image, or simply the content of the image. At the outer limits, it can even be caused by the scanner picking up the grain of the film. (This is mainly a problem with high-speed, coarsely grained film, high-resolution drum and slide scanners, and images that are scaled up significantly.) Sharpening can accentuate those irregularities, resulting in a number of problems—especially with lower-resolution scans.

The most immediately noticeable problem is speckling—usually light pixels in otherwise dark areas. This is especially a problem with low-quality scans that have a lot of noise. You can get rid of speckling to some extent by running a despeckle filter before sharpening.

A less obvious but equally problematic result of sharpening is mottling—often causing people to look like they have bad complexions. It can also destroy the impression of smooth gradations, for instance in a photograph of a sand dune.

FIGURE 21-7 Resolution, sharpening, and aliasing

300 dpi, unsharpened

300 dpi, sharpened

200 dpi, sharpened

150 dpi, sharpened

Higher resolution does little to improve sharpness (though it can improve detail; note the lettering on the ships' bows). Lower resolutions with sharpening applied produce much better results, but aliasing is a danger at lower resolutions, as is visible in the ships' rigging.

CONTENT-RELATED PROBLEMS. There are many other sharpening problems that are content-related, and that you need to watch for on an image-by-image basis. For example, sharpening can accentuate the "highlights" in someone's hair, giving them a case of premature gray. And, in heavy doses, it can give the impression of buck teeth. Watch out for embarrassing accentuation; it's the stuff that caricatures are made of.

Unsharp Masking in Photoshop

Okay, let's get to the juice of the chapter: how Photoshop's Unsharp Mask filter works, and how best to use it. Photoshop offers the most complete unsharp masking controls we've seen on the desktop, so by explaining those controls, we can give detailed suggestions on how to control the sharpening process.

Some other programs (Picture Publisher, for instance) offer similar controls, so you can apply these lessons to those programs, but no desktop program that we've seen offers the range of controls that Photoshop does. You'll need to adjust these suggestions to the limitations of your tools at hand.

Photoshop's Unsharp Mask filter has three variables: Amount, Radius, and Threshold (see Figure 21-8). Each increases or decreases sharpening in different areas and situations. As you're reading through this section, refer to Figure 21-9 at the end of the chapter, which shows the same image sharpened with many different combinations of settings.

Radius. The first step in properly using unsharp masking is figuring out the Radius value. Once you've set Radius, you can go on to adjust the other settings relative to that.

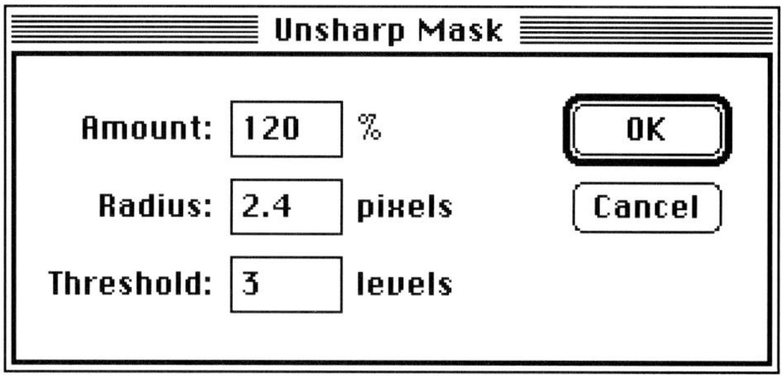

Figure 21-8 Photoshop's Unsharp Mask dialog box

Unsharp Mask

Amount: 120 %

Radius: 2.4 pixels

Threshold: 3 levels

OK

Cancel

The Radius setting controls how wide the blips are on each side of an edge, in sample points. Large values make for wide blips (big halos, lots of sample points involved); small values make smaller blips. The size of the sharpening-induced halo affects how sharp an image looks (larger radius means more sharpening). And in excess, it's the prime culprit in most sharpening-related problems.

A Radius value of 1.0 results in blips two samples wide (four samples total, for the whole light-and-dark cycle). Don't ask us why; we determined this empirically. Just remember that you multiply the Radius setting by four to figure out how wide the halos are.

So what Radius value should you set? Resolution divided by 200. For a 200-spi scan, use a Radius setting of 1.0. For a 300-spi scan, use a Radius setting of 1.5. This results in a halo of 1⁄50 of an inch—1⁄100 of an inch for each blip. It's big enough to provide the sharpening effect, but not so big as to give the impression of artificiality that big halos produce.

Overly large Radius settings are the prime culprits in producing the ugly oversharpened look that stands out so markedly, so watch out if you increase this setting much. The images with a Radius of 2.0 in Figure 21-9 make this clear.

The one situation where you might want to use a larger Radius setting is for large images that people will never look at up close—billboards and (some) posters. In this case, you may need a larger Radius to make the sharpness really pop at the intended viewing distance (a hundredth of an inch ain't much at 100 feet). In that case, try this formula to determine an appropriate Radius setting:

Viewing Distance (inches) × Resolution (spi) × .0004

This formula is based on some theories about the eye's sensitivity to cycles of amplitude modulation over a given viewing arc (we love throwing around big words like that). We haven't had the chance to test this on a billboard, so we suggest you use it as a starting point, and with a pinch (if not a bag) of salt. You might try a small (tabloid-sized) area first, and place it far away to see how it looks.

AMOUNT. We like to think of the Amount setting as the volume, or amplitude, control. It adjusts how intense the blips are on each side of a tonal shift—how much the tonal differences are accentuated. Large values make for big blips (large tonal differences where edges meet); small values make for less significant blips. We're not talking about the width of the halos here—just the *amount* of tonal difference between adjacent samples.

Despite hours of analyzing different results, as well as several messages to the creators of Photoshop, we haven't figured out exactly what the arithmetic of this control is. We have, however, gotten a handle on what values work well. If you use the settings described above for Radius, an Amount setting of 200 percent is a good choice, or at least a good starting point. If the Amount value is set too high, you can get the kind of artificial look we talked about above. It's not much of a problem if you keep the Radius setting low, but if you increase both Radius *and* Amount, images go weird very quickly.

Also be aware that as you increase the Amount setting, you start to reach a point of diminishing returns. The blips around big tonal shifts end up going all the way to white and to black (a problem in itself, because the white blips in particular show up as noticeable artifacts), so increasing the amount has no further effect in those areas. Since those areas of large tonal shifts are the ones you most want to accentuate to give the effect of sharpness, and since you can't accentuate them beyond black and white, increasing the Amount setting beyond a certain point is fruitless.

Large values in the Amount field can also accentuate the problems with noise and mottling that we mentioned earlier, but you can avoid these problems by using the Threshold value, which we discuss next.

THRESHOLD. The threshold value specifies how far apart adjacent tonal values have to be (on a scale of 0 to 255) before the filter does anything to them. If Threshold is set to three, for instance, and adjacent sample points are of values 122 and 124 (a difference of two), they're unaffected by sharpening. So low-contrast areas—those with smooth gradations—aren't affected; the gradations stay smooth.

The Threshold setting is the key to avoiding the mottling, speckling, and artifact-related problems that sharpening can cause. It causes the filter to ignore those slightly-out-of-place pixels, rather than accentuating them.

Low Threshold values result in an overall sharper-looking image (because fewer areas are excluded). High values result in less sharpening. We recommend a setting of three or four for most images. Settings of 10 and above aren't advisable, because they exclude so many areas as to reduce sharpening to near invisibility.

Running the Numbers

We've always been incredibly frustrated by the documentation of Photoshop's Unsharp Mask filter (which, like most of the Photoshop docs, consists of something like this: "Open the dialog box, type in the numbers you want, and click OK.") Finally figuring out what those settings actually *do* (by slogging through endless trial and error—on screen and on output) has been a pain, but having simple formulas we can use for sharpening makes our lives much simpler. We hope it will do the same for you.

Figure 21-9
Various settings for Photoshop's Unsharp Mask filter

200-spi image
133-lpi screen

Unsharpened

Optimal sharpening:
Amount: 200 Radius: 1 Threshold: 3

FIGURE 21-9
Various settings for Photoshop's Unsharp Mask filter, continued

200-spi image
133-lpi screen

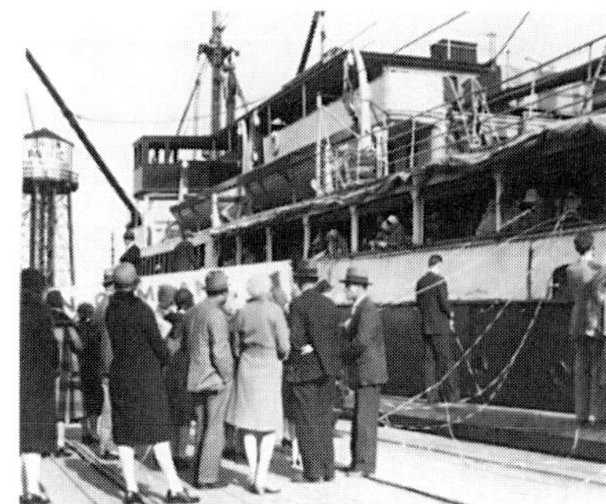

Amount: 75 Radius: 2 Threshold: 3

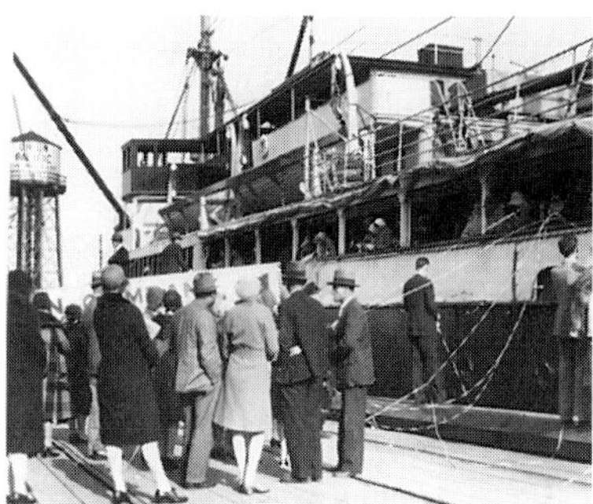

Amount: 125 Radius: 1 Threshold: 3

Amount: 100 Radius: 2 Threshold: 3

Amount: 150 Radius: 1 Threshold: 3

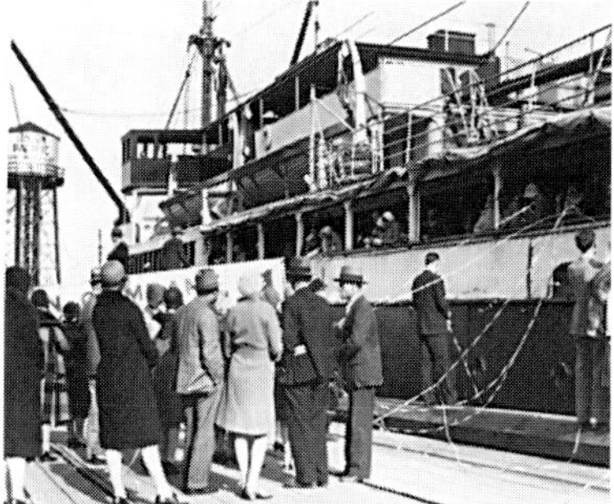

Amount: 125 Radius: 2 Threshold: 3

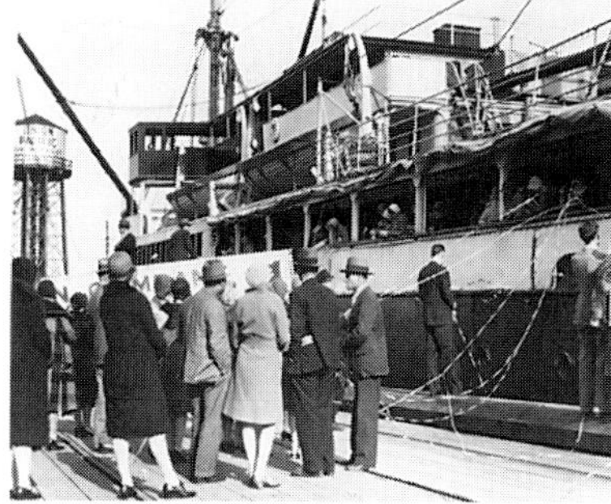

Amount: 175 Radius: 1 Threshold: 3

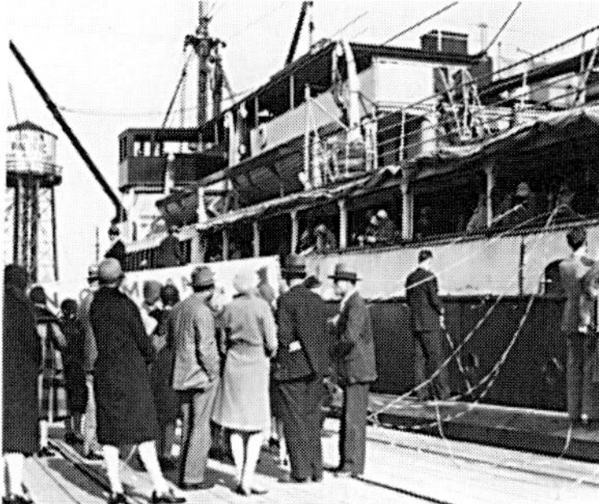

Amount: 150 Radius: 2 Threshold: 3

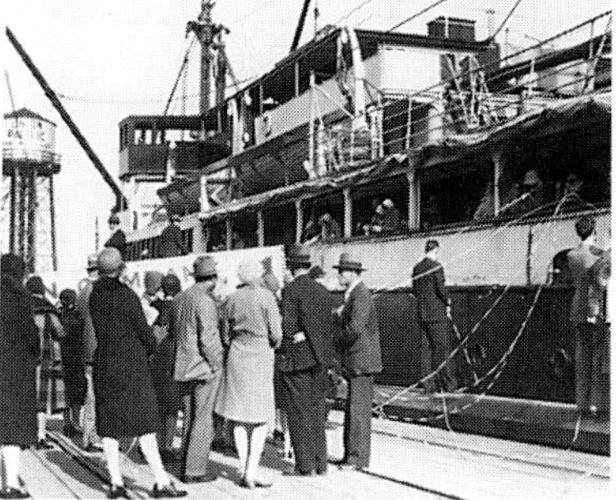

Amount: 200 Radius: 1 Threshold: 3

FIGURE 21-9
Various settings for Photoshop's Unsharp Mask filter, continued

150-spi image
100-lpi screen

Amount: 150 Radius: .7 Threshold: 3

Amount: 200 Radius: .4 Threshold: 3

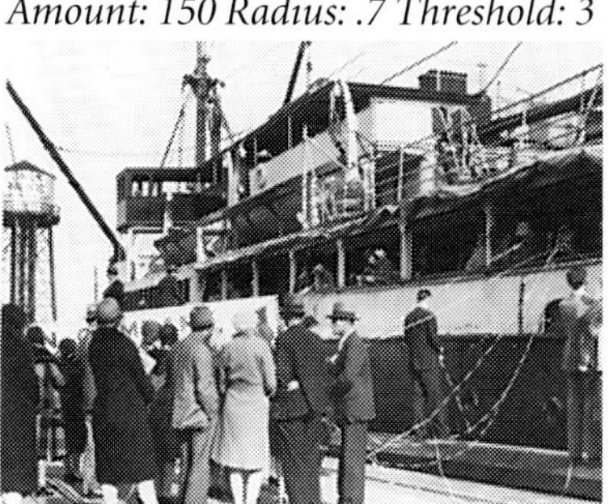

Amount: 180 Radius: .7 Threshold: 3

Amount: 240 Radius: .4 Threshold: 3

Amount: 210 Radius: .7 Threshold: 3

Amount: 270 Radius: .4 Threshold: 3

Amount: 240 Radius: .7 Threshold: 3

Amount: 300 Radius: .4 Threshold: 3

FIGURE 21-9
Various settings for Photoshop's Unsharp Mask filter, continued

150-spi image
75-lpi screen

Amount: 150 Radius: .7 Threshold: 3

Amount: 210 Radius: .4 Threshold: 3

Amount: 180 Radius: .7 Threshold: 3

Amount: 240 Radius: .4 Threshold: 3

Amount: 210 Radius: .7 Threshold: 3

Amount: 270 Radius: .4 Threshold: 3

Amount: 240 Radius: .7 Threshold: 3

Amount: 300 Radius: .4 Threshold: 3

22 Getting Small

Compressing Scanned Images

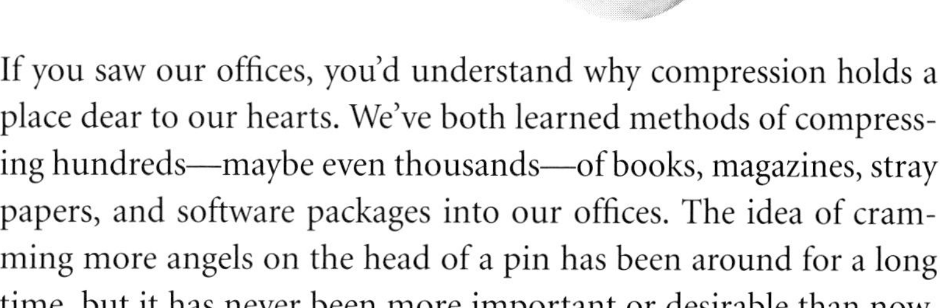

If you saw our offices, you'd understand why compression holds a place dear to our hearts. We've both learned methods of compressing hundreds—maybe even thousands—of books, magazines, stray papers, and software packages into our offices. The idea of cramming more angels on the head of a pin has been around for a long time, but it has never been more important or desirable than now, as people are working with multimegabyte images and limited hard-drive space.

In this chapter we provide an overview of the compression world: what it is, what it does, and generally how it does it. We'll also give you some guidelines for choosing between compression methods, depending on what you're doing and how you're doing it.

Lossy Versus Lossless

When you're playing the game Twenty Questions, what's the first question you ask? "Animal, vegetable, or mineral," right? When you're playing the compression game, the first question you have to ask is "lossy or lossless?"

Lossless

When you compress then decompress an image using *lossless* compression, the file that results contains all of the information of the original. No information is lost, or changed in any way. It's like a kitchen sponge: you can take the sponge and squeeze it down really small, and when you let go it reverts to its original form.

One of the most common forms of lossless compression for images is Run Length Encoding (RLE), which is automatically used for Macintosh PICT images. RLE works by looking for common sample points in a row. For example, let's say you have a black-and-white bitmapped image of a cow. A program that compresses with RLE sees that the first 400 sample points in the image are white, followed by 80 black points, followed by 16 white points, and so on. The program can then compress that data by just recording those values: "400 white, 80 black, 16 white," etc. (See Figure 22-1). This compresses a simple image down to almost nothing, and a complex image just a little (the more complex the image, the less the compression). Fax machines also use RLE.

FIGURE 22-1
Run length encoding

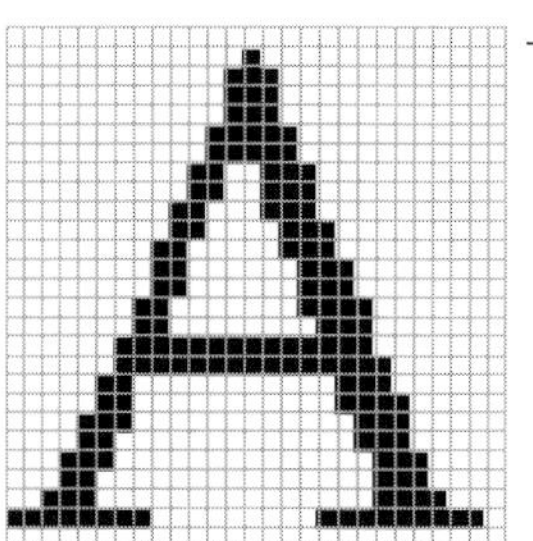

40 sample points of white
1 sample point of black
25 sample points of white
3 sample points of black
25 sample points of white
3 sample points of black
and so on...

RLE is lossless because when you decompress the file, you get exactly the same image as you had before. No information is lost.

Other lossless schemes include LZW (Lempel-Ziv-Welch, though you really don't need to know that) and Huffman Encoding. These work by tokenizing common strings of data. In normal words, that means instead of just looking for a string of the same color over multiple sample points, these methods actually look for trends. If RLE sees "010101," it can't do any compression. But LZW and Huffman are smart enough algorithms to spot the trend of alternating characters, and thereby compress that information.

LOSSY

The second type of compression—*lossy*—actually loses information in the process, but in a very intelligent manner. It's like tightly folding a cotton shirt into a suitcase. If you really cram it in, you might save some room in your suitcase, but when you unfold the shirt, it doesn't look as good as it did before you compressed it.

By losing some information, you can increase compression immensely. Where an LZW-compressed TIFF might be 40 percent of original size, a file saved with lossy compression can be *two percent or less* of the original file size. Lossy compression methods typically give you a choice of how tight you want to pack the data. With low compression you get larger files and better quality. High compression yields lower quality and smaller files.

How much quality do you lose? It depends on the level of compression, the resolution of the image, and the content of the image. Color page D shows an image with various levels of lossy compression from Photoshop. Lower quality means that there are more artifacts—such as grainy sample points—in the image.

The most common lossy compression scheme is currently JPEG (for Joint Photographic Experts Group), though different programs implement the method differently, with varying results. Another method is fractal compression, which is potentially a better system but can take substantially longer to compress (and isn't widely supported). Lossy compression is based on mathematical algorithms that are too complicated to include here (that's a nice way of saying we don't understand the math well enough ourselves). However, one thing you should know is that these are asymmetrical compression methods—it usually takes much longer to compress an image than it does to decompress it.

JPEG++ is a slightly different format that only products from Storm Technology can write, and that all programs that support JPEG can read. It lets you selectively avoid compressing parts of an image, while doing normal JPEG compression to the rest of the image. The could be helpful, for instance, if you had to send an image of a medical X-ray over a telephone line. You could compress everything but the most important parts of the image at a high level, and compress the important parts at a low level.

Here are a few things to remember when working with JPEG. First, note that images with hard, high contrast, and angular areas are most susceptible to develop artifacts from JPEG compression. For example, a yellow square on a green background in a lower-resolution image would look pretty miserable after compression. Second, compressing and decompressing images repeatedly can make images worse than just doing it once (this is reportedly a problem specifically with QuickTime's JPEG compression, but not with most others). Also, if your image is somewhat grainy or impressionistic already, lossy compression probably won't hurt much at all.

NOUN, VERB, ADVERB, OR ADJECTIVE

We don't know who came up with this horrific aberration of the English language, but we just have to stop here and give him or her a piece of our minds. Even though there are no such words as "lossy" or "lossless," it appears that it's all the rage to toss them about. But let's please stem the tide somewhere, okay? Let's not start bandying about sentences such as "Did you compress it lossily or losslessly?" Or, heaven forbid, "Loss, the result of lossyness, can be lost by losing lossisity." And please, please, don't go out and "lossy" your files.

Where's the Compression?

The second part of understanding compression, after lossless versus lossy, is whether the compression is part of the file format itself or whether it's an external archival compression. Let's take a look at each of these methods and when you'd want to use them.

ARCHIVING

If you've ever worked with StuffIt, Disk Doubler, Compact Pro, or PKZIP, you've worked with archival compression. These are all utilities that let you compress and decompress files using lossless compression schemes. What you get when you compress a file, such as an image, is another file that *includes* the compressed image. You can't do anything with that image until you use the archiving program to *extract* the file, creating a new, uncompressed file on disk. Archiving programs have the advantage that you can compress multiple files of all different types into a single archive.

Some compression programs can read other compression programs' files (Compact Pro can read some StuffIt archives, for example), but a program such as QuarkXPress or PageMaker can't read one of those archives. You have to extract the file first (though some compression programs can automatically decompress files when you need to use them).

Archiving programs are great for all sorts of things: when a file can't fit onto a floppy disk, you need to send the file via a modem, or when you're storing files on a disk and you won't be needing them very often. For day-to-day use of image files, however, compressed file formats are where it's at.

COMPRESSED FILE FORMATS

In Chapter 17, *File Formats*, we mentioned that with some formats, you can save a file in compressed form. Unlike archive compression, with compressed files you don't actually need to decompress the file before using it. You can open the file with an image editor, or place it in a page-layout program. The program can read the compressed format, decompressing the file on the fly.

The tradeoff with compressed file formats is speed. Every time you open or place a compressed file, your software needs to decompress the image, which can take some time. And with page-layout programs, the software may need to decompress the file every time you print. But you don't have to decompress the file manually, which is even slower.

Here's a rundown of the main compressed-image formats.

TIFF. The TIFF file specification includes the option of two levels of compression, both lossless. Almost every program out there can read compressed TIFFs (in either compression mode), and almost every image-manipulation program can save compressed TIFFs (though only one program we know of—PageMaker—offers the option of saving TIFFs in your choice of compression modes, either high or medium compression).

PICT. The PICT format on the Macintosh includes RLE compression as a built-in feature of bitmap-only PICT files. As we noted earlier, this method can be great for some images and inconsequential for others. If you're running QuickTime and software that supports it, you can also save PICTs using QuickTime's JPEG compressor. To read those files, you need to be running QuickTime and a QuickTime-enabled program.

ENCAPSULATED POSTSCRIPT. Let's start out this EPS section with a great, simple tip for compression. If your software lets you choose between binary and ASCII encoding for EPS files, choose binary. Bitmaps encoded in binary form are half the size of those encoded with ASCII. You may need to use ASCII encoding if you're passing a file over odd networks or printing using a serial cable, but in most cases, binary is the hot tip.

Now for real compression. Encapsulated PostScript files have not, until recently, included compression schemes. However, PostScript Level 2 is changing that. First of all, EPS images can now be saved in a JPEG compressed format from programs like Photoshop. Because EPS files are totally encapsulated (in theory, another program can't and doesn't need to "get inside" them), page layout programs don't need to know anything about the JPEG format to place them on a page.

The problem turns up when you try to print that image. Your page layout software simply passes the JPEG-compressed EPS through to the printer, and—more likely than not—the printer won't understand it. Remember, the image has to get decompressed to its original state *somewhere* in the process.

Enter PostScript Level 2 from Adobe Systems. Newer laser printers and some imagesetters can decompress JPEG-, RLE-, and LZW-compressed EPS images on the fly at print time (except we don't know of any program that creates RLE-or LZW-compressed EPS images, so you're really just stuck with JPEG). The great benefit of this is that you not only keep the image small on your hard drive, but also you're sending a smaller file down the network lines to your printer, reducing transmission time. The downside is that you can't print the file except on Level 2-equipped devices, and you suffer a performance hit on the output end.

If you need to print a JPEG-compressed EPS on a non-Level 2 printer, try opening the file in Photoshop (it might work) and saving it in uncompressed format. Or try DeBabelizer on the Macintosh. This might save your tofu one late night on deadline.

Small is Beautiful

Remember that the whole idea of compression is to make a file on your hard disk take up less space. Although there are many ways to do this, note that there's always the possibility of simply getting a larger hard drive. Compression always takes extra time somewhere (while compressing and decompressing), and that can be a hassle. Small is beautiful (hey, no comments about our height), but fast and efficient is next to godliness.

23 Just Read It

Optical Character Recognition

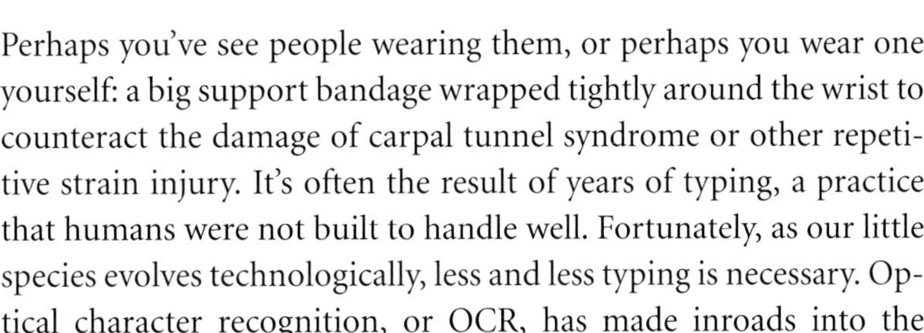

Perhaps you've see people wearing them, or perhaps you wear one yourself: a big support bandage wrapped tightly around the wrist to counteract the damage of carpal tunnel syndrome or other repetitive strain injury. It's often the result of years of typing, a practice that humans were not built to handle well. Fortunately, as our little species evolves technologically, less and less typing is necessary. Optical character recognition, or OCR, has made inroads into the massive task of retyping existing words.

OCR is the technique of scanning text into a computer not as a graphic, but actually as letters, words, and sentences. Sounds simple, doesn't it? It turns out to be a very difficult task, and it's only recently that it's been a usable solution. In this chapter we take a quick romp through OCR, how it works, and what you should be aware of if you want to do it.

How it Do What it Do

Why is OCR so difficult? Can't the computer just see the words? The answer, in a word, is No. Human beings have specialized parts of the brain that identify textures, shape, color, and distance as separate characteristics; these systems help us to look at a tree and know it's a tree, or look at a page of type and read it. A computer, however, can't distinguish between a scan of a page of text and a scan of a photograph. It's all just a bunch of little dots, or samples, saved in the computer as zeros and ones.

However, specialized OCR programs can look through a scanned page and *recognize* characters on it, one letter at a time. There are two techniques for doing this: pattern matching and feature extraction.

PATTERN MATCHING

OCR programs that use pattern matching compare each character on the page to those it has stored in memory. If it has a large library of stored characters, there's a reasonable chance that the program will choose a correct match, but it might take a while to check each letter. Because of the many slight differences in what a character may look like when scanned, pattern-matching OCR programs typically have several versions of what a character should look like.

The problem with pattern matching is that it doesn't handle multiple or unusual fonts very well, because the program has one library of matching characters for each font. If you're OCRing a page of Courier, you first load the Courier library. If you use that library with a page of Helvetica, the program gets very confused and doesn't find many matched characters.

If the OCR program doesn't have a library for the font you're using, you can create one by *training* the program. Programs handle this process in various ways, but typically you can show it a page of text, then tell it what the text really is, and the program will start to learn what that particular font looks like, character by character. This process is training by rote, so you have to keep adding characters for each special case.

FEATURE EXTRACTION

Another method some programs use for OCR is *feature extraction.* In this system, each character gets broken down into various features of the character: diagonal lines, horizontal lines, curves, and so on. The program then matches these features to its understanding of what characters look like. If it sees two vertical lines connected by a horizontal bar in the middle, it figures that it's probably looking at an "H."

The nice thing about feature extraction is that it can recognize a number of different fonts because their basic makeup is so similar. That is, an "s" in one font is pretty similar to an "s" in another font. However, throw in a font with a really bizarre form, and the program chokes. And there's usually no way to train these types of programs.

Additional Features

No matter what sort of algorithm your OCR program uses, there are always times when it can't figure out what character it's looking at (see Figure 23-1). Some programs have additional features that aid them in the quest. For example, just as word processors can spell-check words, some OCR programs can spell-check on the fly. A program might think a word it sees is "loy." But when it can't find that in its dictionary, it reconsiders, deciding that it must be "log" instead (both "y" and "g" have decenders and look similar, and the dictionary might push the program's decision in one direction or another).

FIGURE 23-1
OCR has difficulty with similar characters

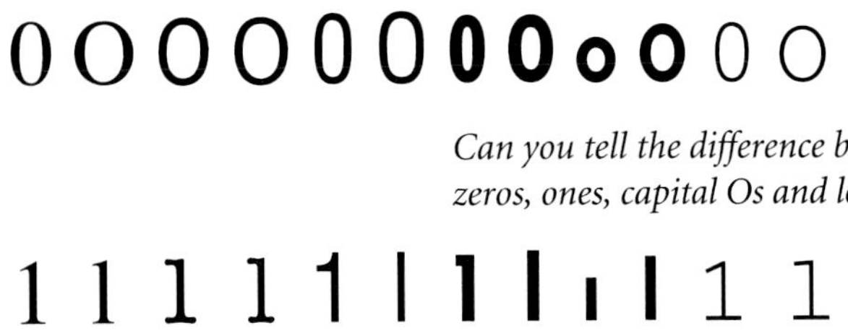

0 O O O 0 0 0 0 o O 0 O

Can you tell the difference between the zeros, ones, capital Os and lower-case Ls?

1 1 1 1 1 l 1 l l l 1 1

Similarly, some programs use context checking. For example, if a letter suddenly appears in the middle of a bunch of numbers, the program might reconsider its choice. Or, if the program thinks it's seeing a "ck" at the beginning of a word, it might be trained to know that it's more likely that the two letters are "ch."

The Last Resort

If your OCR program really can't figure out what a character is, it'll probably just give you a weird character instead. After you process your text scan, you'll almost always have to go back and look for those weird characters, changing them to what letters they should be. Usually, the better the software, the cleaner the scan and the more clear the font on the page, the fewer mishaps (we call them hiccups) you'll encounter.

One thing you should be careful of is what resolution you're scanning at. A low-resolution scan makes it difficult for the program to recognize the characters. But a high-resolution scan makes the OCR unbearably slow. One school of thought recommends 200 spi as a good medium value.

A Few Last Words

Before we move on to other scanning and halftoning issues, we think it's important to cover a few tips and tricks on OCR.

Spaced fonts. Some OCR programs let you specify whether the font on your page is proportional or monospace. The difference? In proportionally spaced fonts, each character has a unique width depending on how wide the letter is. Almost all the typefaces that you see in books, magazines, and newspapers are proportional. On the other hand, typewriters create fonts that are monospaced; each character is exactly the same width. Monospaced characters are significantly easier for OCR programs to read than proportionally spaced characters.

Windowing/zoning. Picture a typical magazine page. There are perhaps a few columns, a photograph, maybe an illustration, and a heading. If you scan that page and process it with an OCR program, the program not only has to figure out where and what the text is, but it needs to figure out in what order the text flows, and so on. Some programs are very smart at this and some are . . . well, let's be polite and just say they *aren't* so smart.

The key is in creating zones or windows on the scanned page that the program can follow. One zone might be a graphic image that can be left alone, another zone might be the first column of text, then another column, and so on. Again, some programs can create these zones for you, and others let you do it yourself.

FORMATTING. If the text you're OCRing is formatted—such as with bold, italic, underline, first-line indent, and so on—your OCR program may or may not be able to figure that out. Therefore, you may either lose the formatting altogether or the program may not even be able to figure out what those letters are. An italic "a" is very different from a normal "a," and a "y" with an underscore looks pretty weird to an OCR program. If you have to do this sort of scanning, make sure your program can handle it well.

STRAIGHT LINES. Many OCR programs have a lot of difficulty with text that has been scanned at an angle. In crooked scans, feature-extraction programs think that horizontal lines are diagonal lines, and pattern-matching programs just go nuts. Scanning the page straight is important to getting good results. This can be a problem, especially with hand scanners. However, there are tools you can purchase or make that help you keep the scanner and page parallel with one another. You can also rotate images in software (and programs such as Ofoto have autostraighten features), but image quality generally degrades with rotation, making character recognition more difficult.

TRANSLUCENT PAPER. If you try to scan and OCR a page on semi-translucent paper, such as newsprint, you may have a helluva' time. Characters from the back side show through and cause all sorts of havoc. One way you can sometimes get around this is by placing a piece of black paper behind the page when you scan it. Then, you can also bump up the scan's contrast somewhat to make the faint back-characters fade away (see Chapter 20, *Tonal Correction*).

The Last Word

Obviously, we can't teach you all you need to know to be a great OCRer here in this book. Each program is slightly different, and many of the techniques rely on program-specific features. However, what we do want you to come away with from this chapter is the understanding of what's required to translate thousands (or millions) of zeros and ones into readable, editable, forgettable text. And the next time you're faced with typing 127 pages of someone's manuscript into the computer, you'll stop and think about wristbands and scanners.

24 Photo CD

Bright Lights, Big Images

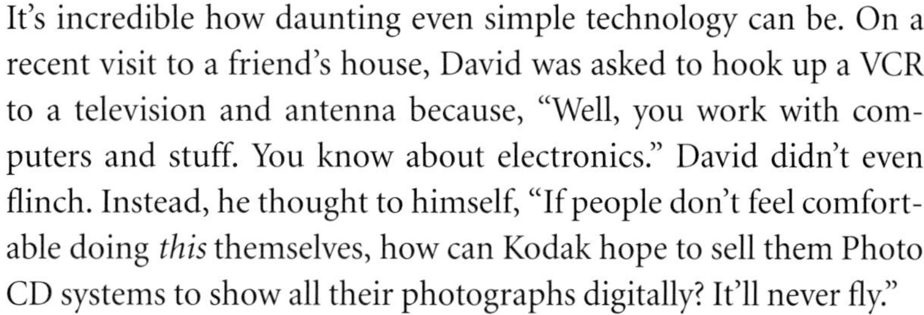

It's incredible how daunting even simple technology can be. On a recent visit to a friend's house, David was asked to hook up a VCR to a television and antenna because, "Well, you work with computers and stuff. You know about electronics." David didn't even flinch. Instead, he thought to himself, "If people don't feel comfortable doing *this* themselves, how can Kodak hope to sell them Photo CD systems to show all their photographs digitally? It'll never fly."

He was right. Photo CD, a technology for quickly and easily putting photographs on a CD-ROM, while heavily marketed to consumers, is not relevant for that market yet. However, it's very relevant to desktop publishers. In this chapter we're going to look at why it's becoming important and how you can use it to cut time and money from some areas of your production work.

Photo CD Basics

No matter what the sales guy tells you, Photo CD is not a revolutionary concept. It's simply a new format for storing images on disk. You already know that you can scan photographs and save them as TIFFs, or EPSes, or whatever. But Photo CD can save you time, money, and hassles. Photo CD images typically come on CD-ROM disks, but they don't have to. Once you've gotten a Photo CD image, you can copy it to your hard drive or a removable cartridge.

As electronic publishers have found over the past year, Photo CD isn't just an expensive, high-ticket, goofy consumer toy; it's also a high-quality tool for photographers, designers, desktop publishers, and prepress folks. We'll start out by looking at how Photo CDs are made, then discuss the quality and technical issues of the captured images, move on to how you can access the images, and finally look at the cost and time savings involved.

Making a Photo CD

You have to give Kodak credit for getting Photo CD into both high-end service bureaus *and* the corner K-Mart. The big workstation that creates Photo CDs is called a Photo CD Imaging Workstation (PIW). Kodak first sold these to huge photofinishers that do pickup and developing for drugstore chains and other retailers. That means that you can get Photo CD images from many different outlets.

And because the process of making a Photo CD is the same whether it's done in massive numbers in Louisville or painstakingly at an imaging center in Los Angeles, chances are you're going to get a decent scan to disk no matter where you are. (Of course, it's more likely that a professional photo lab or a prepress service bureau will make great scans.)

Film

At the time of this writing, all Photo CD images are made from 35mm film. However, Kodak just released a scanner and new software to handle larger formats that professional photographers use, such as four-by-five-inch transparencies. Disks made using this system are called Pro Photo CDs.

You can supply either slide or print film (positive or negative) to be scanned, but it's cheaper to supply film that hasn't been developed yet and that hasn't been cut into strips. If you supply cut film, each strip has to be identified (see below) and fed into the machine separately; long rolls of film can be fed in much more quickly.

IDENTIFICATION

The PIW operator has to look at the film before scanning and identify the manufacturer, type, and speed. This information is printed along the edge of all film, repeating every frame. For example, "Kodak 5096 Gold 200" means that it's Kodak film, Gold is the type, 5096 is its unique ID, and 200 is the film's speed. All contemporary film also has a bar code on the edge which has that information encoded in it.

Identifying these values is a critical step, because different film emulsions—the photosensitive coatings that receive the image—have different color characteristics. Kodak has analyzed their own film as well as many from Fuji, Agfa, 3M, and Ilford, and rolled those findings into what they call declarations—descriptions of their color characteristics.

If the film you're using isn't "declared," there's a good chance that your scan will be poor. But if you go to a service-oriented place to get the scan, they may be able to build a custom declaration for your film.

SCANNING AND WRITING

The PIW workstation—a roaring-fast Sun Sparcstation—does an 18Mb scan of each frame in about 10 seconds. The scan is processed into a Photo CD file, or *image pack.* Those image packs are written onto a CD-ROM.

The last stage in the process is printing a little color thumbnail proof of each image. These get ganged up on a small sheet and slipped in with the disk. These thumbnails are so small that you can't really see many details, but it's better than having no proof at all.

WHAT YOU GET

The images on a Photo CD aren't stored in an easy-to-access format, like TIFF. They're stored in an image pack, which is a proprietary format that Kodak fiercely guards—for now. That means that you may not be able to just stick the disk in and start reading images off it. You need to know how the images are stored in each image pack and have the right software to grab them.

STORAGE

On the standard Photo CD format, called Photo CD Master, you can store approximately 100 color images, or about 150 grayscale images (made from so-called black-and-white film). Each image is encoded into a single image pack. They call it a pack because it actually contains five different resolutions of the file (see Table 24-1). You can figure out which one to use based on the two-times-screen-frequency formula we discuss in Chapter 19, *Choosing Resolution.* The last two columns of the table suggest output sizes for different resolutions at different frequencies using a more liberal 1.5-to-1 resolution-to-frequency ratio.

TABLE 24-1 Image pack resolutions for Photo CD Master with output sizes

				Output size at	
Real name	*Goofy name*	*Dimensions*	*Size*	*133 lpi*	*150 lpi*
Base*16	Poster	2,048 × 3,072	18Mb	10 × 15.5	9 × 13.5
Base*4	Large	1,024 × 1536	4.5Mb	5 × 7.5	4.5 × 7
Base	Snapshot	512 × 768	1.1Mb	2.5 × 3.8	2.3 × 3.5
Base/4	Wallet	256 × 384	288K	1.3 × 2	1.1 × 1.7
Base/16	Thumbnail	128 × 192	72K	Not applicable	

In some of the software you use to access Photo CD images, Kodak assigned strangely undescriptive names to each size (like poster or wallet), but these goofy names are meaningless. As always, what's really important is the dimensions of the image, because with the number of pixels and the size of the image you can determine resolution (see Chapter 16, *Scanned Images*).

The "real" names are based on how Photo CD is used in the consumer world (or more accurately, *not* used). The Base image fits on a regular TV set, which can display between 400 and 500 pixels vertically. The others are defined up or down from that.

Pro Photo CD will let you use images at Base*64, or four times the size of Photo CD Master. One of these images is 72 megabytes.

Typically, you won't need this kind of image unless you're doing large-format, high-frequency images such as tabloid magazine covers; you could easily print a full 11-by-17-inch page with a bleed at 150 lpi using the Base*64 image. Or you might need it if you need to capture a very small detail of a larger picture, such as one face in the middle of a college class photo.

COLOR FIDELITY

It's important to understand that the Base*4 and Base*16 images are compressed using a lossy compression scheme (see Chapter 22, *Getting Small*). So color fidelity is an issue when you're using those resolutions. Here's how Photo CD's compression scheme works.

The human eye perceives variations in luminosity—differences from light to dark—more readily than differences in color. So Photo CD uses luminosity as its main component; it stores one channel of luminosity and two color channels which more or less represent the red and blue channels. Then, instead of remembering all three channels of the image at every resolution, Photo CD retains all the luminosity information and throws away almost 60 percent of the color information for the two highest resolutions.

That means that the two highest resolutions in each image pack (Base*4 and Base*16) contain progressively less color information. This sounds kind of dangerous, but the way Kodak does it is very clever. The biggest problem is that colors don't show as much variation and can seem flat in the high-resolution images.

If you *really* care about color fidelity in an image, you should be careful about using the higher resolutions from Photo CD. We're not going to say don't use them, because we've seen some really spectacular images come out of the Photo CD system, but this drawback is something to be aware of.

ADJUSTING FOR OUTPUT

All the issues we've discussed throughout this section of the book, such as sharpening and tonal correction, apply to Photo CD images as well. A Photo CD image is similar to a raw, uncorrected scan you'd get off a good scanner. However, there are a few differences.

The Photo CD workstation handles the overall tonal balance for you. Because Photo CD is currently geared for TV display, the image's saturation is boosted, and the tonal range is shifted to look good on a television screen. You may need to compensate by decreasing saturation a bit, and adjusting the tonal range to compensate.

Photo CD images are not sharpened at all, so you'll need to take that step before placing images in your publications.

In 1994, you can expect to see Print Photo CD, which will use the offset press color gamut rather than a television's. With this system, each scan on your Photo CD will come as an image pack and as a CMYK TIFF at full resolution. Each scan will also have a file containing information about the separation for use with Scitex, Crosfield, and Linotype-Hell high-end systems.

SINGLE AND MULTISESSION

Just because you can put over 100 images on a disk doesn't mean you have to wait until you have that many images to make a Photo CD. If you only put a single roll of 36 exposures on a disk, you can go back and put more on later. That's because the PIW can write to the disk on more than one occasion or *session.*

Note that the overhead for each session's directories takes up space on the disk—so much space, in fact, that each session decreases your storage by about five images.

There's one other problem with multisession disks, however: you need a multisession CD-ROM drive to read anything past the first session. Almost every drive sold in 1992 or earlier is a single-session drive only (though some of them can be upgraded to support multisession disks). On the other hand, almost every drive sold in 1993 or later supports multiple sessions. The Apple CD150 is a single-session drive; the CD300 is multi-session. If you're uncertain about yours, you can call Kodak at (800) 242-2424, extension 36, to get a list of multisession Photo CD-compliant drives.

ACCESSING IMAGES

As we said, Kodak uses a proprietary technique to store images on a disk. However, they license a toolkit for software developers to incorporate support for Photo CD within applications or utilities. Most major applications provide some degree of support for accessing Photo CD images; plus, Apple released a system software update that provides simple Photo CD access.

PHOTOSHOP

Fortunately, Photoshop was one of the first programs to support Photo CD. Although you can retrieve Photo CD images directly from the Open or Open As dialog boxes in version 2.5, we suggest using the Kodak Photo CD Acquire Module instead, which can be purchased directly from Kodak. While the Open dialog box lets you open a picture at any of the saved resolutions and into the grayscale, RGB, or LAB color mode (See Figure 24-1), the plug-in gives you even more possibilities (See Figure 24-2).

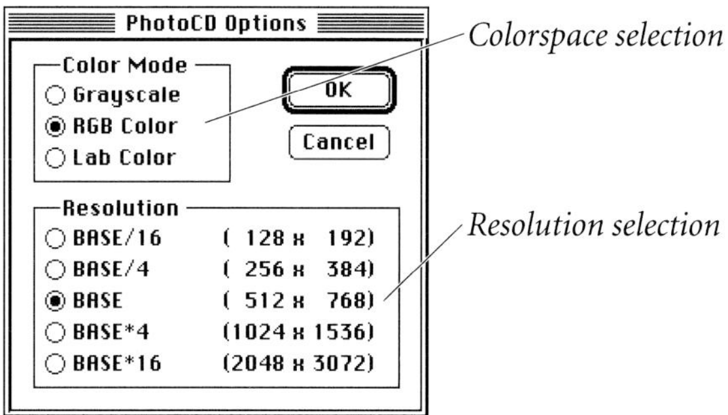

FIGURE 24-1 Photoshop Open dialog box for Photo CD images.

The two most important additional features in the acquire module are the ability to select just a portion of the image rather than the whole thing, and the ability to make some gamma and color correction before acquiring. You can use slider controls to change color cast, saturation, and overall brightness, or select a stored setting from a pop-up menu. Adobe includes a similar module with Photoshop that allows images to be acquired directly into Kodak's color-management system.

FIGURE 24-2 Kodak Photo CD Acquire Module for Photoshop

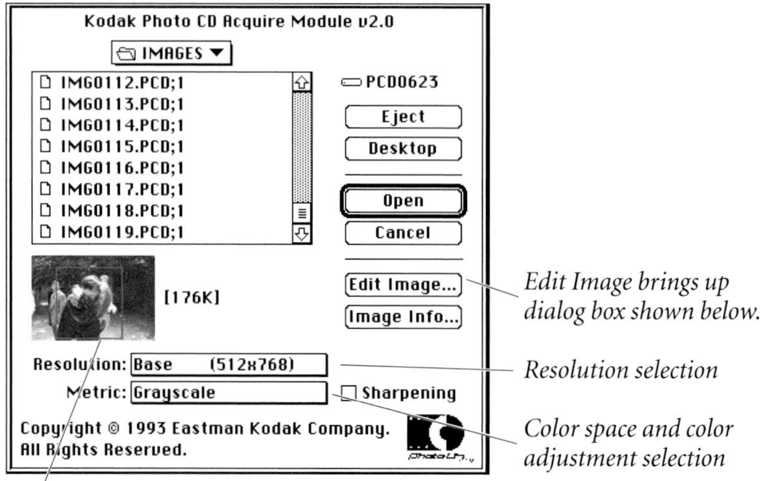

Preview can be cropped; size reflects cropped image

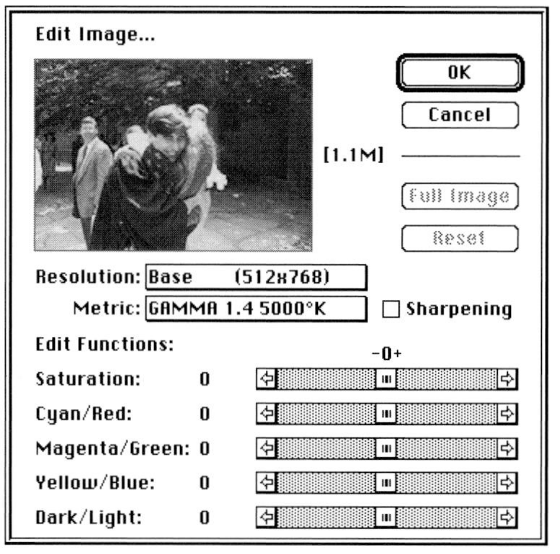

The Edit Image dialog box has a larger preview for viewing corrections more accurately.

The sliders allow adjustment of color cast, saturation, and brightness; the preview is updated dynamically.

When opening an image from the Open dialog box and deciding what color space to open a Photo CD image in, take into consideration what sort of color transforms you're planning on making. If you are planning on converting the image into CMYK mode, for instance, then it's better to open the image in LAB (a perceptual color space that maps pretty directly from Photo CD's native Photo YCC space). That way, the image doesn't have to be transformed into RGB and then into CMYK. The fewer transformations, the more likely that colors will remain true.

QUARKXPRESS

The latest version of QuarkXPress (3.2) comes with an XTension that lets you import Photo CD images directly into picture boxes. The images get imported as RGB files, and if you're using the EfiColor XTension that also comes with QuarkXPress version 3.2 (or later), you can make the RGB-to-CMYK separations yourself. The problem is that the Photo CD XTension doesn't give you any control over how the image is imported, as the Acquire modele in Photoshop does. In fact, you can't even tell QuarkXPress which resolution you want to import; it determines that at output time.

We typically leave the Photo CD XTension alone or put it away someplace else. Instead, if we want to import a Photo CD image, we import it as a PICT file. We talk about this method below.

PAGEMAKER

PageMaker has a slightly different approach to importing Photo CD images than QuarkXPress's. You select the image through the Place dialog box, and it automatically brings up a filter with some options (see Figure 24-3). You can import the image as either grayscale or RGB at any of the five resolutions. But you can also let PageMaker decide the color space and resolution based on the values you've entered for screen frequency and output device. If you let it decide on resolution, PageMaker will substitute the correct resolution when you print your document.

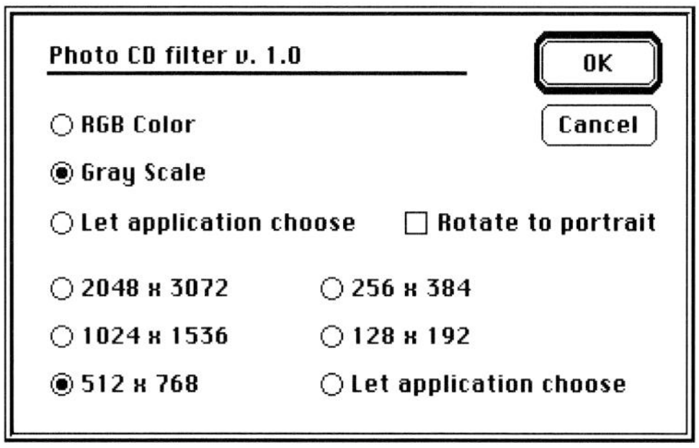

FIGURE 24-3
PageMaker's Photo CD filter

Photo CD filter v. 1.0

OK

Cancel

○ RGB Color

◉ Gray Scale

○ Let application choose

☐ Rotate to portrait

○ 2048 x 3072	○ 256 x 384
○ 1024 x 1536	○ 128 x 192
◉ 512 x 768	○ Let application choose

Just as with Quark, however, it's senseless to import a file unless you can make tonal adjustments and sharpen it. So while we recommend this technique for mock-ups and previews, it's not wise for final output work.

MACINTOSH SYSTEM 7

Apple and Kodak do a lot of business together, so it's natural that Apple would wholeheartedly back Photo CD. However, this is one instance where business politics have actually made it *easier* to use a technology. The key is using a Macintosh with at least System 7.0, QuickTime 1.5, and the CD-ROM 4.0.1 drivers (of course, all these version numbers are minimum configurations; if you're using later versions, like System 7.1 or QuickTime 1.6.1, it works even faster.)

If you're set up with all that, then all sorts of cool things happen.

THUMBNAILS. First, when you insert a Photo CD disk, you get a progress dialog that tells you the system is making Photo CD color icons. When it's done, you have a folder in the Photo CD's window on the desktop, called Photos. This folder, in turn, contains five other folders labeled with the five picture resolutions described earlier in this chapter. If you open one of those folders, you can see thumbnails of the images right in the Finder (see Figure 24-4).

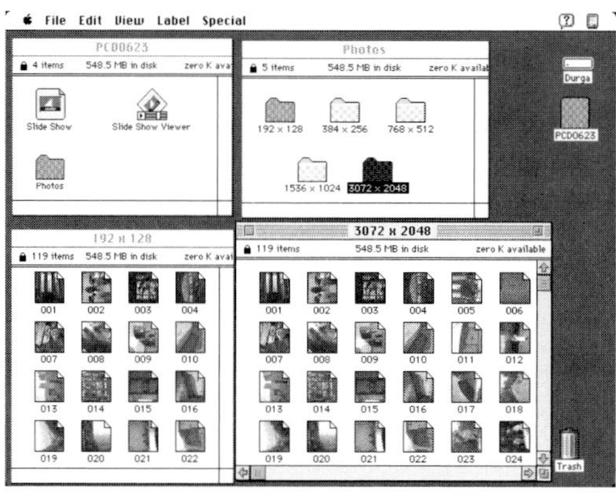

FIGURE 24-4 Photo CD in the Macintosh Finder with QuickTime and the new CD-ROM software

SLIDE SHOW. The next cool thing that you'll notice is the Slide Show file and application. Like the Photos folder, these don't really exist on the disk; but the Macintosh shows them to you just as though they did. If you double-click on the Slide Show, a window appears and lets you run through the images one at a time by clicking on buttons (see Figure 24-5). The Slide Show works just like the standard QuickTime Movie Player, except it uses still images.

FIGURE 24-5
Slide Show

Basic Movie Player controls: play/pause, slide to specific image, fast rewind, and fast forward.

You can also load the image you're looking at in Slide Show into a program automatically using the View or View As items from the menus, or by double-clicking on the image.

APPLICATION SUPPORT. Perhaps the greatest feature of having the newer versions of system software and QuickTime is that they let you open Photo CD files from within any application. Even if the program doesn't inherently support Photo CD itself, it can open the Photo CD images because the system fools the program into thinking that it's looking at PICT files. This trick also lets you drag any file from the Photos folder to another disk. The Finder acquires just that resolution and copies the file to your disk with a preview icon.

Once again, however, we typically like to open the image in Photoshop first and make our modifications. Then we save it as a TIFF or EPS image before we bring it into another program. But just to know that you *can* bring a picture in directly is a good thing in our books.

PHOTO CD ACCESS

Kodak sells a little utility called Photo CD Access (for the Macintosh and the PC) which is so inexpensive ($40) that we need to at least touch on it here. If you don't use Photoshop or another Photo CD-aware program, you can use Photo CD Access to preview, open, rotate, and crop your Photo CD images. The biggest drawback is that it requires a lot of RAM to work (the entire image has to fit into available memory).

Why Photo CD Is a Good Thing

So finally we get to the good stuff: how Photo CD can make your life happier and your business more prosperous. Photo CD technology has a number of advantages over other forms of scanning, including desktop scans or even drum scans. With Photo CD you can get scans that approach the quality you get from drum scanners, but for only a couple of bucks an image. You also save money on storage.

Cost and quality. Photo CD scans cost a *lot* less than any comparable scan. You could easily pay between $10 and 30 for a typical CCD scan, or $40 to $200 for a drum scan. On the other hand, one high-end, professional photofinisher we know of in Boston charges $5 per image for a calibrated and tweaked Photo CD scan. And you can get very good Photo CD scans for $2–3. Some photofinishers are now charging as little as 90 cents per image.

If the quality of a $2 scan is almost as good as a $100 scan, which would you prefer? As we noted earlier, you start to lose color fidelity in the higher-resolution images. But in many cases, the loss of color isn't noticable on an offset press. See Color page C for a comparison of drum scanning versus Photo CD.

COST OF STORAGE. You effectively get about 1.8 gigabytes of data on a Photo CD. That's the equivalent of 40 SyQuest 44Mb cartridges, or over 1,200 high-density floppy disks. The charge for a Photo CD disk may be about $15. Or, if you scan more than 24 images at once, the disk itself—the media—may be thrown in for free.

If you could store 100 images on $2,000 of storage media or on $15 worth, which would you choose?

SPEED OF ACCESS. One other great thing about Photo CD is that the technology stores a number of resolutions in one image pack. That means that you can open the image at just the resolution you need. If you only want a little preview or for-position-only image, you can import one at Base/16. That's much faster than opening the full-page high-resolution scan, and then downsampling to the size you need.

Also, because the images are stored together and can be quickly previewed, you save a lot of time in reviewing images for use. You can even use image database programs such as Imspace's Kudo, Aldus Fetch, or Kodak's Shoebox to keep track of multiple CDs.

NETWORK ACCESS. Because of the way Photo CD images are stored, you can access the pictures very quickly over a network. To see a thumbnail of a TIFF, for instance, your computer has to pull over the entire file. With Photo CD, it can pull over just a low-resolution image. Then, you can acquire the compressed, high-resolution image and let it decompress on your computer, saving even more time. Now, if you put six of Kodak's 50-disk jukeboxes (all SCSI-chained together) on your network, you could have more than 540,000 megabytes of images at your fingertips . . .

What, Me Worry?

If you're like us, when anyone talks for this long about how good something is, you get nervous. Well, it's true, Photo CD is not the answer to all life's questions. It may not even be the answer to your image needs. But it msay just prove to have enough utility and quality to make it another weapon in your desktop publishing arsenal.

PART 3
POSTSCRIPT

25 PostScript Halftoning

setscreen, setgray, image, imagemask, and settransfer

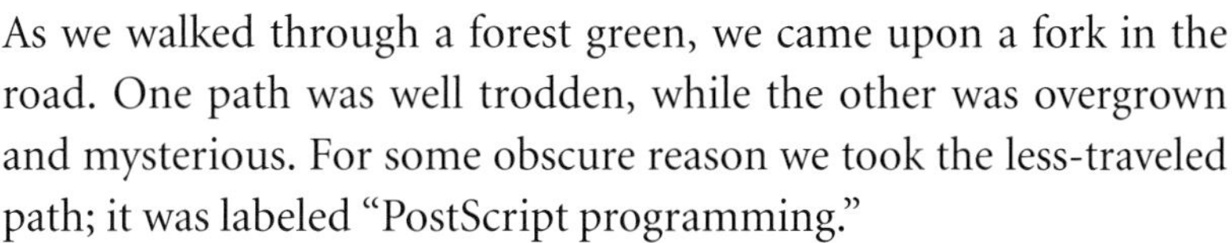

As we walked through a forest green, we came upon a fork in the road. One path was well trodden, while the other was overgrown and mysterious. For some obscure reason we took the less-traveled path; it was labeled “PostScript programming.”

If you’ve taken a similar path, and know something about the bizarre world of PostScript, this chapter and the next may make some semblance of sense to you. On the other hand, if you’ve taken the well-worn path (believe us, you’re not alone, and with good reason), you probably don’t know anything about PostScript—or want to—and you can just skip these chapters.

It is to their credit that John Warnock and Chuck Geschke, when developing PostScript over 10 years ago, saw fit to include a powerful but simple halftone generator in the PostScript engine. Printing grays to non-PostScript printers requires the software on the computer to generate the appropriate halftone screens at the resolution of the printer. For the application to print to a PostScript printer, however, it only needs to send the screen angle, frequency, spot shape, and gray level; hardly a daunting task.

We’re not going to turn you into PostScript programmers in this chapter. We will, however, explain the basic PostScript operators for scanning and halftoning, so you can recognize them and understand their operation when you’re perusing PostScript code. Maybe you're just curious. Maybe, someday, this will save your butt at three in the morning.

There are five primary PostScript operators of importance for scanning and halftones: *setscreen*, *setgray*, *image*, *imagemask*, and *settransfer*. There are others in PostScript Level 2, but they are so little used as of this writing that we don't cover them here. There are also several operators for color screening and imaging (and even more of them in Level 2), but they are largely derivatives of the basic five. If you understand these five, you have the essential tools for mucking around with PostScript halftoning.

Here's a brief rundown of the four operators.

SETSCREEN. This is the basic PostScript halftoning operator. You use it to specify frequency, angle, and spot shape.

SETGRAY. This operator simply specifies a gray level, from 0 to 100 percent.

IMAGE. This operator is used to image all bitmaps.

IMAGEMASK. Only one-bit images can be used with this operator.

SETTRANSFER. This operator adjusts a PostScript device's gray-response, so you can control what actually gets printed when a given gray level is requested.

SETSCREEN

The key to assigning halftone screens, angles, and spot shapes in PostScript is the *setscreen* operator. This operator takes three arguments: the frequency, the angle, and the spot function. While the first two are obvious, the third is a little more complex, so we've written a whole chapter on it (see Chapter 26, *Spot Functions*). A very simple example, which sets a 75-lpi, 45-degree line screen, reads *75 45 {pop} setscreen*. Here's a more real-world example of setscreen in action.

```
currentscreen          % put existing screen arguments on stack
/spot exch def         % save existing spot function in "spot"
pop pop                % throw away frequency and angle
133 45 spot setscreen  % put frequency, angle, and spot function
                       on stack, call setscreen
```

The *currentscreen* operator loads the device's existing screen arguments on to the stack in the same order as *setscreen* requires. In the above code, "spot" becomes a variable defined as the spot function of the existing screen, preserving it for use in our new screen. The frequency and angle settings are thrown (popped) away. Finally, new frequency and angle settings are placed on the stack, followed by the spot function. The *setscreen* operator uses those three arguments, resulting in a 133-lpi, 45-degree screen with the previously existing spot function.

SETGRAY

There are simpler operators than *setgray* in the PostScript lexicon, but not many. To define a gray level for an object in PostScript, you use *setgray* with a value between zero and one (zero is black; one is white). Here's an example that draws a 100-point square, and fills it with a 40-percent tint.

```
100 100 moveto
0 100 rlineto
100 0 rlineto
0 -100 rlineto
closepath
.6 setgray
fill
showpage
```

The *.6 setgray* line sets the gray level, and the *fill* operator fills the current object with that gray.

IMAGE

The *image* operator, which is used to image most bitmaps, is somewhat more complex than most PostScript operators. Rather than providing a uniform gray fill, *image* goes through each sample point in a string of data, setting the gray level for each sample.

The *image* operator takes five arguments: the width and height of the bitmap, the number of bits per sample point, the image matrix (the relationship between the image's coordinate system and PostScript's), and the data acquisition procedure (the method for reading the ensuing image data). What you get is an image in a one-unit square at the coordinates 0,0 (a unit is one point large until you scale it up). Here's *image* in action:

```
/width 20 def                 % Number of sample points horizontal
/height 19 def                % Number of sample points vertical
/BitsPerSample 8 def          % Bits per sample point
/TheMatrix [width 0 0 height neg 0 height] def
                              % Image matrix
/picstr 20 string def         % Convert width to string and save in picstr
/DataRead {currentfile picstr readhexstring pop} def
                              % Data reading procedure
100 100 translate             % move to right place on page
200 190 scale                 % scale the image up from a one-unit square
width height BitsPerSample TheMatrix /DataRead load image

DFDDDDE0E1DEE1E8E0D0D0DAE4E1DBDCDAD3D4D9DDDCDBDDDCD9C08B5C313056
96BBDFE0DEDBE0D3D5D7D6D4D8C94911190D0D10213C93DCDDDCE1CED0D4D4D6
C3643A576F687167716C439DDDD5D7D0D3D6D2D8B131617A91AEB7A590855342
CDD5D7D9D9D6D4D7C24A61738EA4A5A18A7E5F3AA6D8D7D7D8D7D0D5B35C3A33
346A8F5E404645479AD3D3CFCAD3CFD0AF772217241C281D230F245BA2C9D2C9
C2C4C9C8BA96575766658D69735D4D65BDCCD2CBCDD0D2C9C2BA5C6467658E68
777A68A5D0CCD1CACBC8C2A97A543854603D3B50686960B8D1CECCC6D0C38E4C
16041E3B665A63637B5F2F3096C9C9C6CCC19C672B0C3739367481804F453B38
87BDC6C3C9C9BB9151293C9B552A3340678A4B75AEC0C1BBD1CFCEBB916746B9
CB684D96CD9073A6BDC1C2BDD3D6D4D2C6AC8CBFA353A9DDC59AADC2C9C5C3C2
D4D3D4D6D7CFBBC4938ED0D4BDB7C9CCCCC7C1BDD5D1CFD3D7DAC9C7B0BFD0D2
C8C9D0D1CDC6C2BFD8D5D3CAC4C1BFC0B1B7C0BFC0C3C5D3CEC5C1C1

showpage
```

Note that if the image is saved in binary rather than hexadecimal format, you need to use *readstring* rather than *readhexstring.*

IMAGEMASK

While it's similar to *image, imagemask* serves a somewhat different purpose, and it only works with black-and-white (one-bit) bitmaps. The *imagemask* operator uses a one-bit image to set up a mask through which "paint" can be applied to the page. It takes the same width, height, image matrix, and data acquisition parameters as *image.* But instead of the bits-per-sample value (the image is expected to be one bit deep), it takes a true/false boolean argument that tells it which areas to paint. True paints the black areas of the bitmap; false paints the white areas.

SETTRANSFER

The *setgray* and *image* operators tell the PostScript device what gray level to spit out. But what if the actual, printed gray level isn't what you want? What if you (or your software) ask for a 15-percent gray, and you get 20 percent instead? It's a common scenario: the output from laser printers and imagesetters varies based on many factors.

Given that variation, there's obviously a need for some overriding operator that controls how a device images grays, without the need to go through a whole PostScript program changing all the *setgray* and *image* values. That metaoperator is *settransfer.* Its job is to map the grays requested to a set of grays that PostScript uses for imaging.

The *settransfer* operator takes a gray level and converts it into another gray level, depending on the algorithm you give it. For instance, *settransfer* can be used for simple things like producing negatives through the command *{1 exch sub} settransfer.* Or it can be used for strange things, such as in this procedure written by Bill Woodruff, to convert all the values between .25 and .75 to .5, while passing all the others through unaltered: *{dup .25 ge 1 index .75 le and {pop .5} if} settransfer.*

CALIBRATION

The most interesting uses of *settransfer* come when you start correcting the gray curve for the output device you're using. The following bit of code, for instance, adapted from Pat Wood's work in *The PostScript Language Journal*, maps the grays to provide a more even density transition on the Apple LaserWriter at its default frequency. The same method can be used for posterization (stairstepping the gray values for special effects) or for linearizing an imagesetter so a request for 10-percent gray yields a 10-percent tint.

```
/transarray [
0 0 0 0 0 0 0 0 1 1 1 2 2 3 3 3
4 4 5 5 5 6 6 6 7 7 8 8 8 9 9 10
10 10 11 11 12 12 12 13 13 14 14 14 15 15 15 16
16 17 17 17 18 18 20 20 20 22 22 24 24 24 26 26
26 28 28 31 31 31 34 34 37 37 37 40 40 42 42 42
44 44 46 46 46 48 48 48 49 49 51 51 51 52 52 54
54 54 55 55 57 57 57 59 59 60 60 60 62 62 62 63
63 65 65 65 66 66 68 68 68 69 69 71 71 71 72 72
72 72 72 73 73 73 74 74 75 75 75 76 76 77 77 77
78 78 78 78 78 79 79 79 80 80 81 81 81 82 82 83
83 83 83 83 84 84 84 85 85 85 85 85 86 86 86 87
87 87 87 87 88 88 88 88 88 89 89 89 89 89 90 90
90 90 90 91 91 91 91 91 92 92 92 92 92 93 93 93
93 93 94 94 94 94 94 94 95 95 95 95 95 96 96 96
96 96 97 97 97 97 97 97 97 97 97 97 98 98 98 98
98 98 98 98 98 98 99 99 99 99 100 100 100 100 100
] def
{255 mul cvi          %multiply gray times 256 and make integer
transarray exch get   % look up gray value in array
100 div               % return gray setting in 0 to 1 range
} settransfer
```

The array consists of 256 values, one for each of 256 gray values. You could just as easily use 50, 100, or 200 values. The important thing is the gray-scale mapping. Let's say the requested gray level is

.5. The transfer function multiplies it by 255 and converts it to an integer, yielding 127. It looks up the 127th value in the array (72), and divides it by 100, resulting in .72 setgray (a 28-percent tint) when you request .5.

This array is designed to work on LaserWriters and other Canon CX engine printers at their default frequency (nominally 60 lpi, actually 53). If you change the frequency, you will need to change the array as well, since spot variation is more pronounced at higher frequencies (see Chapter 5, *Setting Your Screens*). Figure 25-1 shows two graduated fills from white to black. In the first, *settransfer* is unaltered. The second shows the results with the transfer array above.

FIGURE 25-1
settransfer and grays

Typically, this sort of calibration is extremely time-consuming and tedious; fortunately, in this case other people have done most of the hard work for you. Two programs, Kodak's Precision Imagesetter Linearization Software and Technical Publishing Software's Calibration Software for PostScript Imagesetters, can generate and manage multiple transfer functions so that you can work with multiple calibration settings (one for coated stock, another for newsprint, and so on). They typically have an array like the one in Figure 25-2, into which you can type densitometer readings from sample output.

So you don't really need to know how to program transfer functions, but it can be useful to know what these programs are doing behind the scenes—especially when some program insists on overriding the transfer functions you've set up with one of those calibration programs.

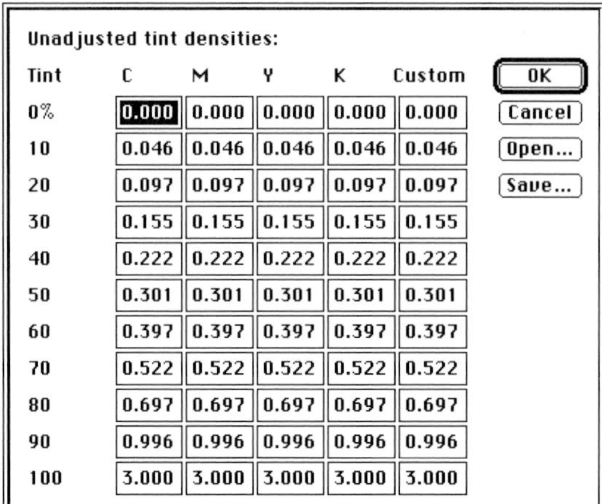
Unadjusted tint densities:

Tint	C	M	Y	K	Custom
0%	0.000	0.000	0.000	0.000	0.000
10	0.046	0.046	0.046	0.046	0.046
20	0.097	0.097	0.097	0.097	0.097
30	0.155	0.155	0.155	0.155	0.155
40	0.222	0.222	0.222	0.222	0.222
50	0.301	0.301	0.301	0.301	0.301
60	0.397	0.397	0.397	0.397	0.397
70	0.522	0.522	0.522	0.522	0.522
80	0.697	0.697	0.697	0.697	0.697
90	0.996	0.996	0.996	0.996	0.996
100	3.000	3.000	3.000	3.000	3.000

FIGURE 25-2
Adobe Separator's Tint Adjustment dialog box

THE WEIRD WORLD OF POSTSCRIPT

PostScript's halftoning engine is extremely powerful, providing you with all the tools you need to create great screens. If you're not already overloaded, then hold tight: the next chapter explores this world a little further with an in-depth look at spot functions.

26 Spot Functions

Inside the Halftone Cell

Now it's time to don your helmet, grab hold of your pick-axe, and play the part of a spelunker—one of those wild-eyed explorers who seem to thrive on the atmosphere of deep, dark, dank caves. But we won't be looking at stalagmites and stalactites today. Rather, we're going deep into the heart of the halftone spot. What makes up these tiny critters? What mysterious algorithm decides a spot's shape?

In Chapter 5, *The Glorious Spot*, we described a variety of spots for different purposes. And in Chapter 25, *PostScript Halftoning*, we quickly passed over the concept of spot functions, saying that they were just part of the setscreen operator's variables. In this chapter we're going to see exactly what makes a spot function tick.

Note that we're talking about normal ol' rational tangent screening here. Many developers have created alternate screening methods (see Chapter 8, *Beyond the Spot*), but those are proprietary methods that we won't discuss in detail here.

The Dot Grid

As soon as you call *setscreen*, PostScript begins to create a perfect halftone cell—something that can be repeated (tiled) over a large area—the whole page if necessary. The cell's shape and size depend on the printer's resolution, the screen angle, and the screen frequency (see Figure 26-1).

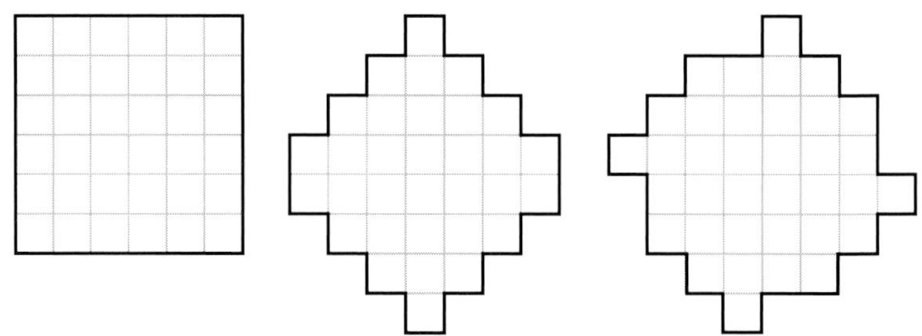

FIGURE 26-1
Various halftone cells

Once the cell shape is determined, the PostScript interpreter can build a halftone spot inside it. To do this, it needs to look at each printer dot that falls inside the cell. Each dot has a particular coordinate value from (-1,-1) to (1,1). The value (0,0) is right in the center (see Figure 26-2).

THE FUNCTION

The spot shape is generated by a tendency for dots to turn on. The interpreter assigns a "tendency value" to each dot in the halftone cell. This value corresponds loosely to gray levels that the halftone cell will be asked to simulate. Tendency values are always between -1 and 1.

For example, at a 25-percent gray level, all the dots in the cell that have a value of -1 to -.5 would turn on. To create a 50-percent tint, all the dots with values up to and including zero would turn on. (The tendency values resulting from a spot function don't actually have to span the whole range from -1 to 1. The could go from -.4 to .6, for instance; the important part is the *relative* ranking of dots within the cell.)

Spot functions work by taking two arguments—the X and Y coordinates of the dot in the cell—and handing back a tendency value for each dot. So the dots are ranked, in the order in which they're to be turned on. When the interpreter receives a gray value, it turns on the first five dots, the first 50 dots, or whatever is necessary to achieve that tint percentage. To make this clearer, let's look at a simple example—a spot function that creates line screens.

FIGURE 26-2
Dot coordinates in a cell

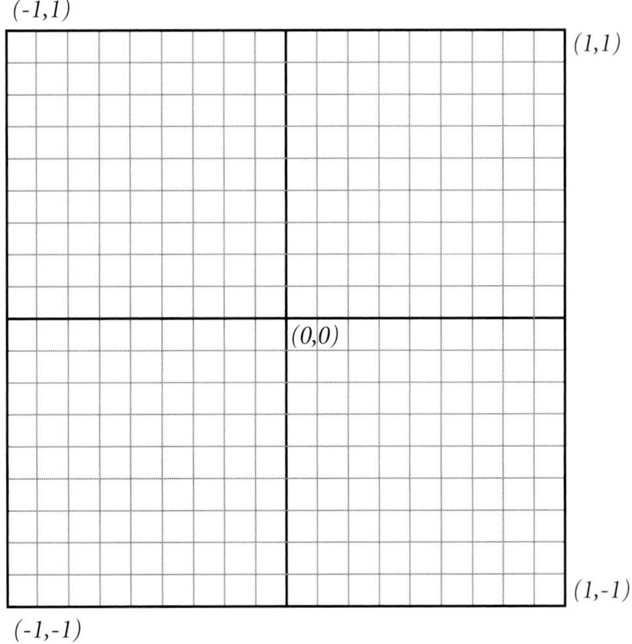

LINE SPOT

Perhaps the simplest spot function is *{pop}*, which creates a line screen. The PostScript interpreter puts the X and Y coordinates of each dot on the stack (so Y is on top, with X next). The line spot function simply throws away the top (Y) value (The *pop* operator removes the top item from the stack).

The result? Each printer dot is assigned a tendency equal to its X value. So all the dots on the left side have low values, and therefore turn on first. All the dots on the right side have higher values, and turn on later—only with darker grays. Figure 26-3 shows a representation of this cell, detailing each dot's coordinates, tendency value, and ranking. The dots always turn on in the order that they're ranked; each dot of a higher ranking is turned on only after all lower ranking dots have been turned on.

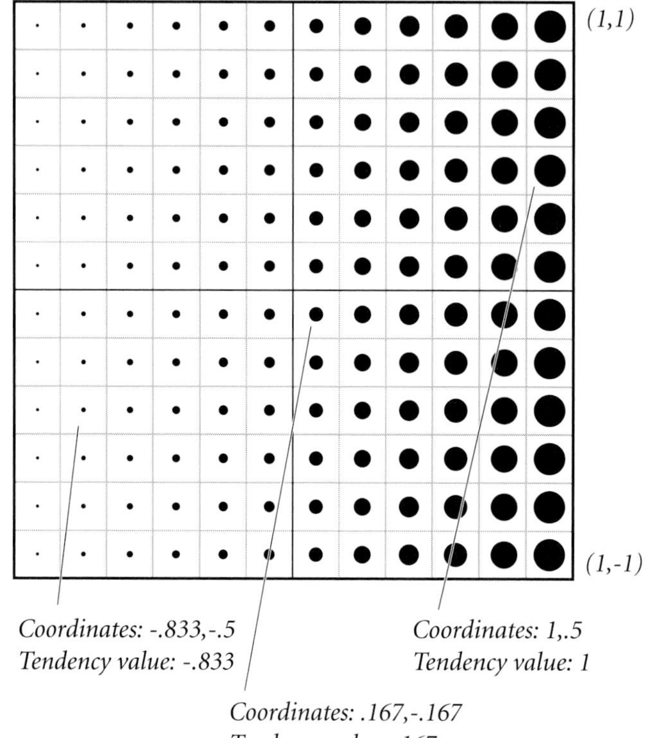

FIGURE 26-3
Line spot

ROUND SPOT

The next spot we'll look at is the normal round spot: *{dup mul exch dup mul add 1 exch sub}*. This is the PostScript equivalent to "$1-(x^2+y^2)$." The result is a round spot that increases in size until it fills the entire cell. Figure 26-4 shows a representation of this cell.

The more common alternate to the standard round spot is *{abs exch abs 2 copy add 1 gt {1 sub dup mul exch 1 sub dup mul add 1 sub} {dup mul exch dup mul add 1 exch sub}ifelse}*. This spot function is conditional: up to a certain point in the cell (corresponding to an approximately 50-percent spot), the normal round spot is used. Beyond that, an inverted spot is used. That is, the spot starts to be built from the corners into the center rather than the center out.

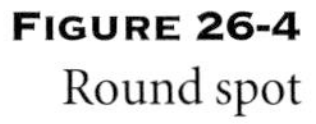

FIGURE 26-4
Round spot

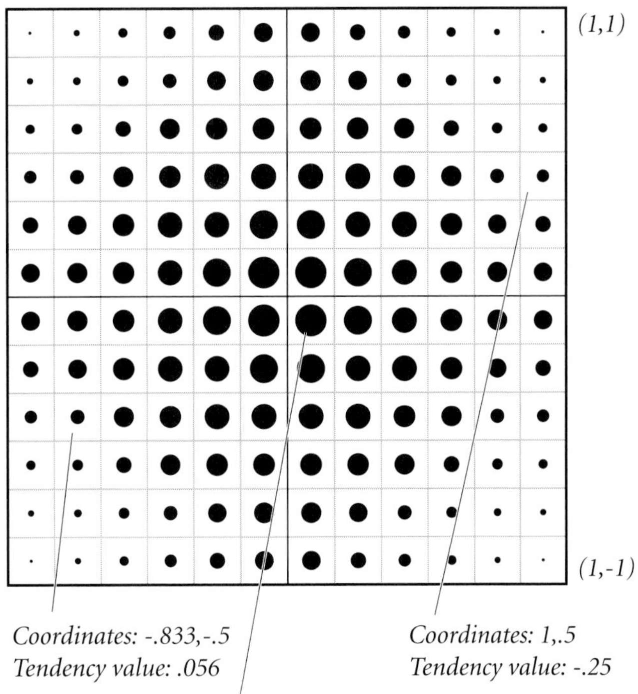

MORE SPOTS

As long as you can come up with a PostScript function for it, you can create spots of almost any shape. Here are some of our favorites (see Figure 26-5). If you're hacking PostScript code, you can type them into a setscreen argument. Or, if you use a program that supports custom screens (as Photoshop does via the Screens button in the Page Setup dialog box), you can type them into a dialog box. With Photoshop, don't type the procedure name (the slash and the word)—just the curly brackets and what's inside them (you can also include the "bind" at the end; it can make the code run faster).

FIGURE 26-5 Some favorite spot functions

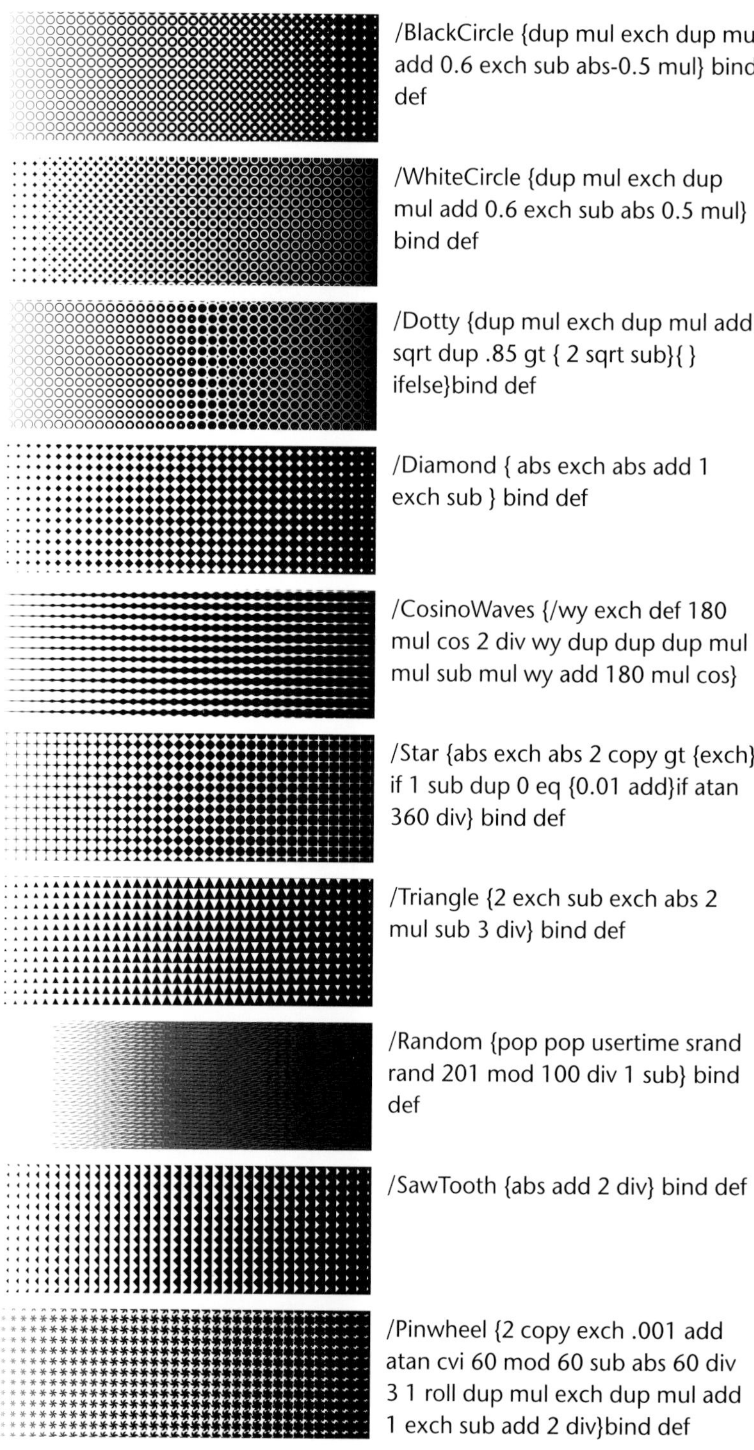

FIGURE 26-5 Some favorite spot functions continued

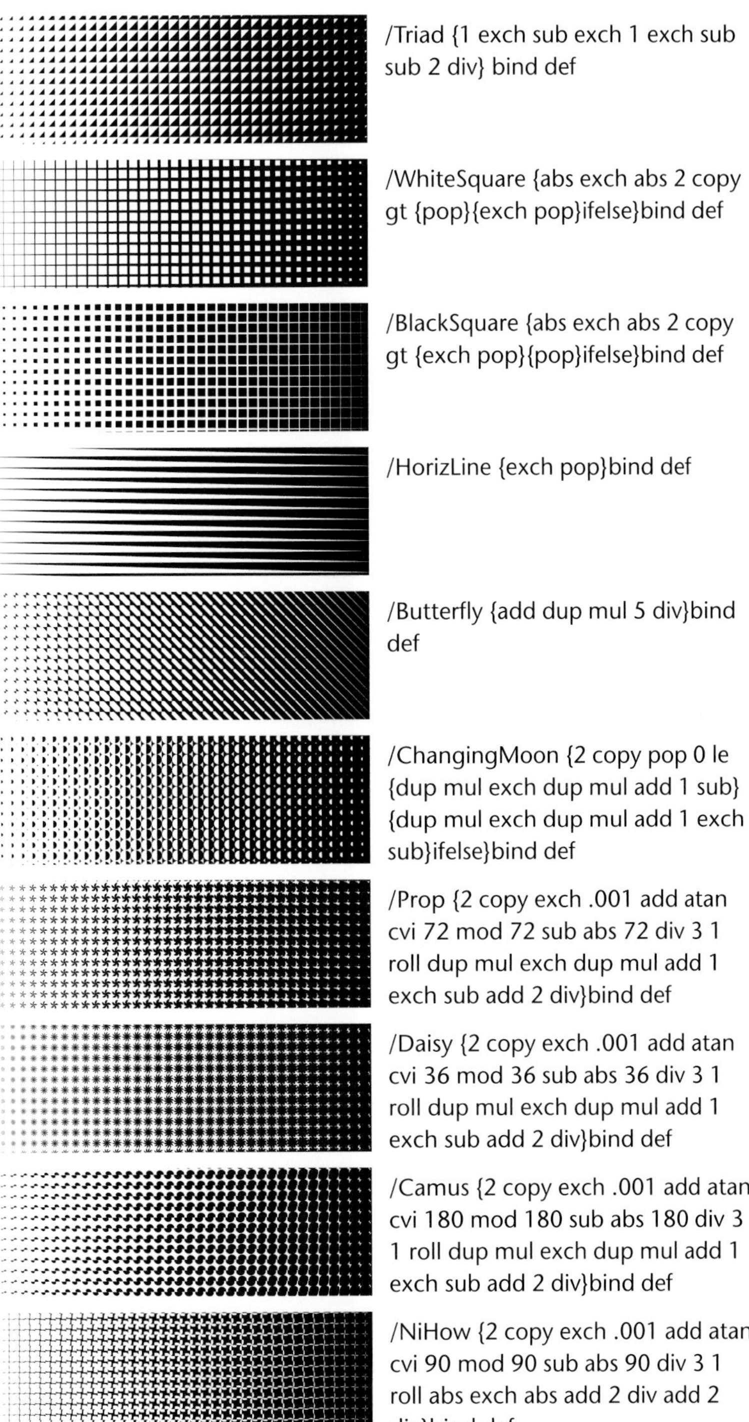

The Strangest Spot Functions

As you can see, spot functions can be pretty strange. Typically, however, you'd never use the weird spot functions except for special effects. Even then, they're a little dodgy to use, and Adobe generally suggests you not use low-frequency custom-spot halftones for special effects. We think they're pretty cool, though. It's a good risk to take.

When it comes to *really* strange spot functions, though, nothing beats these: custom pictures in a spot function. Figure 26-6 shows one of these. The code is too complex to include here, but we wanted to show you that it *can* be done.

Figure 26-6
Really strange spot function

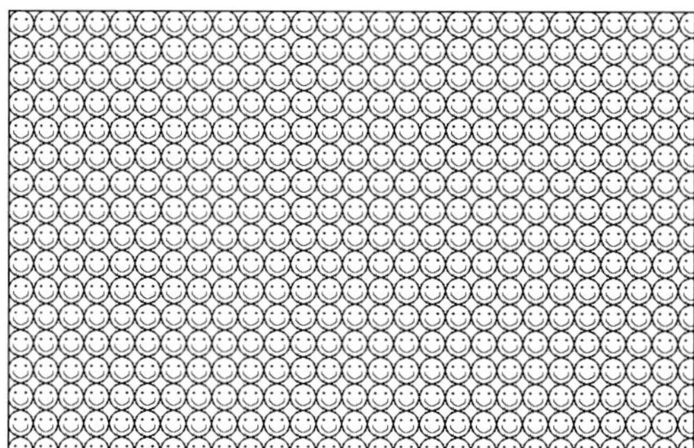

Spot Functions and Page Orientation

One final caveat about using strange spot functions for special effects: you can get messed up if you rotate a page on output. Since they're working with actual device pixels, halftone screens are by necessity working in *device space.* If you rotate *user space*—for instance printing a page in transverse mode to an imagesetter—the orientation of the spots doesn't rotate with the page.

So if you're using triangle spots at a coarse screen frequency to point your reader's eye to the right, watch out when you rotate the page. The triangles will be pointing north or south, not east. The same problem arises with line screens, or any other orientation-dependent, asymmetrical spot function.

Caching Spots

If you really want to understand PostScript's halftoning (and if you're still reading this chapter, you obviously do), you have to understand how it caches spots. When you use *setscreen*, the PostScript interpreter builds every possible spot for that frequency/angle/spot function/resolution combination, and caches them all away. It builds a spot with one dot on, one with two dots on, one with three dots on, up to the total number of dots in the cell.

Then when it's time to actually image a halftone—a gray fill or a scanned image—the PostScript interpreter simply pulls the appropriate spots out of its cache. If it needs a 50-percent gray value, it grabs the spot with half the dots turned on, and plops it down. If it needs a 25-percent spot, it grabs the cached spot with one quarter of the dots turned on.

There are a couple of implications to PostScript's spot-caching behavior. First, it's fast. Once it has built and cached all the spots, the interpreter just uses them; it doesn't have to recalculate the spot function for each individual spot. (This is why irrational screening—covered in Chapter 12, *Frequencies, Angles, and Moirés*—is so processor-intensive. Spots can't be cached; the interpreter must calculate each spot in an image on the fly.)

The second implication of spot caching is the impossibility of true random screening (for instance, a diffusion dither) using PostScript's standard spot-function-based screening. You can write a random spot function easily enough (*pop pop usertime srand rand 201 mod 100 div 1 sub*), and PostScript will build a spot with dots turned on randomly. But since PostScript builds all the spots once and caches them, all the 50-percent spots in an image are the same. And they're essentially the same as the 45-percent spots, but with more dots turned on.

If you print the job a second time, the spots will be different. But in each print job, all the spots of a given gray level are the same; and they're substantially the same as the other spots, but with more or less dots turned on. The result is a patterned look that changes from print job to print job, but doesn't give you the randomized dither look you might be looking for.

On the Spot

We're kind of weird this way, but we really like playing with spot functions. Strange spots are rarely useful for production work (but when you need 'em, you need 'em), and they can be problematic for special design effects, but they sure are fun to play with.

Part 4
Applications

27 Image Applications

Messing with Bitmaps

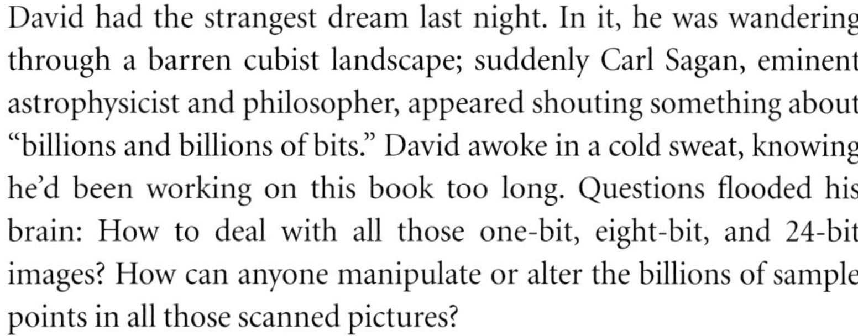

David had the strangest dream last night. In it, he was wandering through a barren cubist landscape; suddenly Carl Sagan, eminent astrophysicist and philosopher, appeared shouting something about "billions and billions of bits." David awoke in a cold sweat, knowing he'd been working on this book too long. Questions flooded his brain: How to deal with all those one-bit, eight-bit, and 24-bit images? How can anyone manipulate or alter the billions of sample points in all those scanned pictures?

Fortunately, he remembered Photoshop and all the other image-manipulation applications on the market. These are designed by trained bit-gurus to handle any kind of bitmapped image. Programs like Photoshop, PhotoStyler, Picture Publisher, Cachet, and DeBabelizer are the stronghold of image manipulation. In this chapter we'll look at each of these and explore their strengths and weaknesses.

Photoshop

Photoshop is *the* central tool for working with scanned images on the desktop, and for that reason we've covered it in quite some detail throughout the book. (We've also used it to create the bulk of the illustrations). In particular, we cover two of Photoshop's most important tools—Levels and the Unsharp Mask filter—in detail in Chapter 20, *Tonal Correction*, and Chapter 21, *A Sharper Image.*

We could write a whole book on Photoshop's other features, but that's already been done. See, in particular, *The Photoshop Wow! Book*; *Four Colors, One Image*; and *Photoshop in Black and White*, all published by—who else?—Peachpit Press. There are a couple of halftone-related features, however, that we want to discuss here.

CONTROLLING SCREENING IN PHOTOSHOP

Photoshop provides better control over screening on output than any other program we've seen. Click the Screens button in the Page Setup dialog box, and up comes the Halftone Screen dialog box (see Figure 27-1).

FIGURE 27-1
Photoshop's Halftone Screen dialog box

Halftone Screen
Halftone Screen
Frequency: 53 lines/inch
Angle: 45 degrees
Shape: Diamond
OK
Cancel
Load...
Save...
Use Accurate Screens
Use Printer's Default Screen

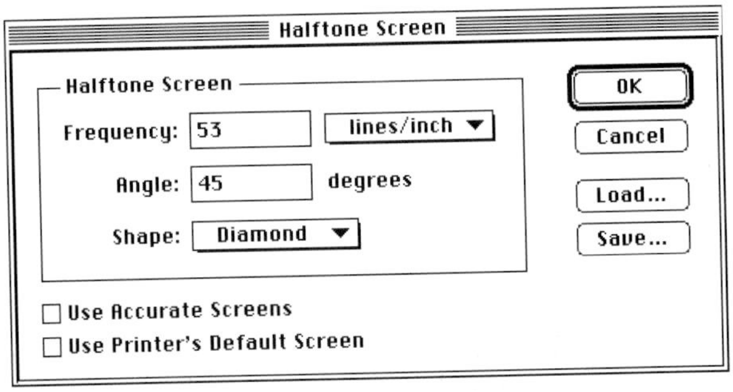

When you specify frequency, angle, and spot shape in this dialog box, you get that halftone every time you print the image (until you change the settings). You can also include this screening information in a saved EPS version of the image by selecting Include Halftone Screen in the Save as EPS dialog box. (For more on controlling screens in EPS files, see Chapter 13, *Controlling Halftone Screens*.)

It's that easy. However, there are three items in the dialog box that deserve a little more discussion: the Shape popup menu, and the Use Accurate Screens and Use Printer's Default Screens checkboxes.

SHAPE. The Shape pop-up menu (see Figure 27-2) lets you specify what halftone spot function you want to use. You can choose among six canned spot functions, or define your own. The Ellipse spot function is especially interesting and relevant, because it generates the transforming elliptical spots that we describe in Chapter 6, *The Glorious Spot.*

FIGURE 27-2
Custom spot shapes in Photoshop

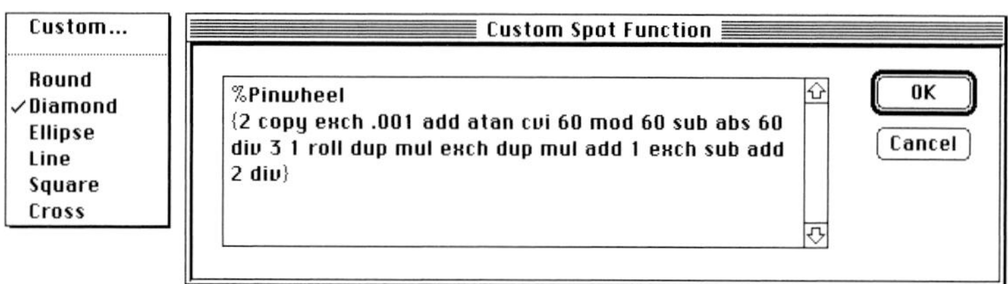

If you want to use a custom spot function, such as the ones we provide in Chapter 26, *Spot Functions*, choose Custom from the Shape pop-up menu. You can then type the PostScript code in the dialog box. Remember to include the curly brackets on either side of the function. Note that custom halftone screens are rarely necessary except for low-frequency, special-effect screens.

USE ACCURATE SCREENS. The Use Accurate Screens checkbox causes imagesetters equipped with Adobe's Accurate Screens technology to use that technology, snapping requested frequencies and angles to ones that are included in the optimized screening sets. The thing is, no one we know really uses Accurate Screens. Most people use screening routines provided by the major imagesetter vendors: Agfa's Balanced Screens, Varityper's ESCOR, or Linotype-Hell's HQS.

USE PRINTER'S DEFAULT SCREEN. Turn the Use Printer's Default Screen checkbox on if you want your image to print using the output device's default screen settings. Note that your service bureau may have changed their imagesetter's default settings, so the defaults you get may be different from what they were when the machine came from the factory.

HALFTONING WITHIN PHOTOSHOP

If Photoshop is remarkable in any way (and it's remarkable in many ways), it's because it actually can do a lot of the things that PostScript can. It can open Illustrator PostScript files and convert them to bitmaps, just like a PostScript interpreter. And it can convert any grayscale file to a black-and-white bitmap, halftoning it in the process.

To turn a grayscale image into a halftone, choose Bitmap (a.k.a. "bilevel bitmap") from the Mode menu. You'll see the dialog box in Figure 27-3. Choose Halftone Screen as the conversion option, and you get the dialog box in Figure 27-4.

In the Halftone dialog box, you can set the frequency and angle of the resulting halftone, and choose among Photoshop's built-in spot shapes. Unfortunately, you can't use a custom spot function in this dialog box. If you want a custom spot function, you might try creating it by selecting Custom Pattern in the Bitmap dialog box (you have to define a pattern first).

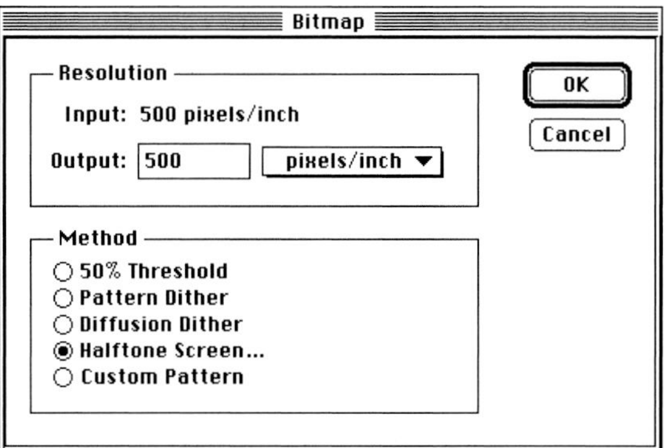

FIGURE 27-3 Photoshop's Bitmap dialog box

FIGURE 27-4 Photoshop's Halftone dialog box

By the way, if you want to play around with your own version of stochastic, or frequency-modulated, screening, try converting grayscale images to black-and-white bitmaps using the Diffusion option. The result is very similar to the techniques available on imagesetters with stochastic screening.

PhotoStyler

Aldus PhotoStyler version 1.1 for Windows machines offers many of the tools and controls that we've discussed throughout the book. It has especially good tonal correction, but unfortunately its sharpening filters leave quite a bit to be desired. (Version 2.0 was announced as this book was going to press; let's hope they've improved it.)

TONAL CORRECTION

The Gray/Color Map control in PhotoStyler operates in pretty much the same way as Photoshop's Levels control—described in Chapter 20, *Tonal Correction*—though as we say in that chapter, PhotoStyler actually offers a bit more information. Not only does it show the histogram and the input and output level controls, it also shows you the gray-map curve as you adjust those controls.

What's missing, however, is the incredibly useful tone-clipping display that Photoshop provides.

SHARPENING

PhotoStyler 1.1 doesn't offer an unsharp mask filter, so you have to monkey around a bit with the filters that do exist, trying to simulate the flexibility of unsharp masking (see Chapter 21, *A Sharper Image*). Remember: you're trying to achieve a sharpening halo of 1/50 of an inch or so (1/100 of an inch for the blip on each side of a tonal change).

Using PhotoStyler's Sharpen or Sharpen More filter is equivalent to using Photoshop's Unsharp Mask filter with a Radius setting of .5—appropriate for images under 150 spi. The difference between the two is in the amount of accentuation between adjacent samples.

Sharpen Heavily is equivalent to using a Radius setting of 1.0 in Photoshop—good for images in the 200-spi range. Since you can't control the parameters explicitly, however, you might find it better to run the Sharpen or Sharpen More filter twice (which increases the radius to 1.0 while significantly increasing the amount of tonal difference) rather than using the Sharpen Heavily filter.

We're hoping that version 2.0 of PhotoStyler will offer a true unsharp mask filter with adjustable parameters—especially for the threshold value, which is totally absent in the current incarnation.

CONTROLLING HALFTONING

Command Central for halftoning in PhotoStyler is the Page Setup dialog box (see Figure 27-5). You can choose to have your PostScript printer do the halftone by clicking the Halftone by Printer checkbox (if you don't understand this, see Chapter 7, *Who Does the Halftone?*).

FIGURE 27-5 PhotoStyler's Page Setup dialog box

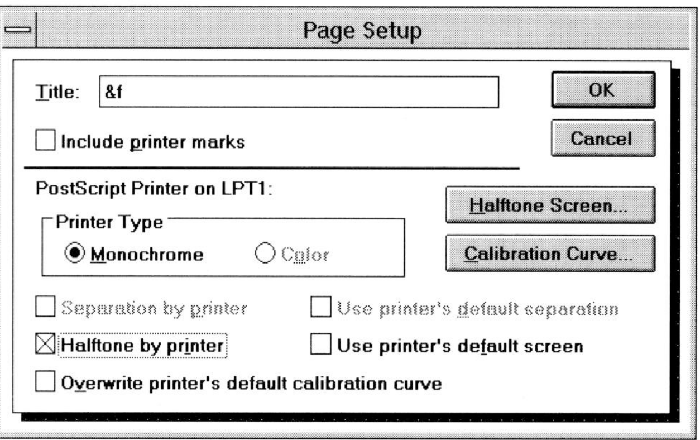

Uncheck the Use Printer's Default Screen checkbox, and the Halftone Screen button comes to life. You can use this button to set the frequency and angle for each of the four process colors, and the spot shape for the whole job. These Page setup controls allow you to control halftoning when printing from PhotoStyler, but you can't save those settings in EPS files.

HALFTONING WITHIN PHOTOSTYLER

You can halftone a grayscale image within PhotoStyler, converting it to a big, black-and-white, halftoned bitmap. With a grayscale image on screen, select Convert to Black & White from the Image menu. You'll see the dialog box in Figure 27-6. Choose a spot shape, frequency, and angle, and PhotoStyler halftones the image for you, right then and there.

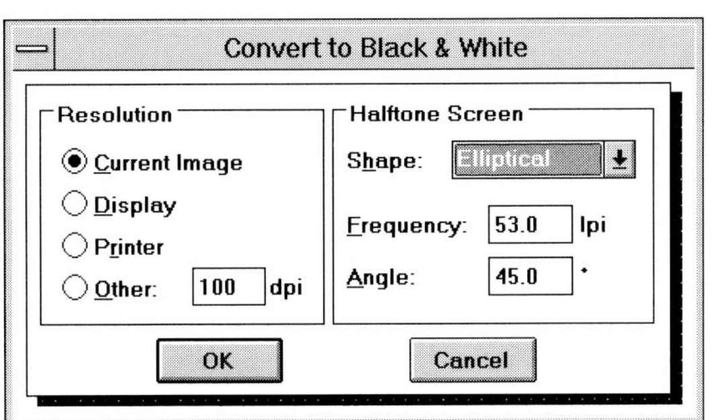

FIGURE 27-6 PhotoStyler's Convert to Black & White dialog box

You can also use the Diffusion "shape" to achieve nontraditional halftoning, and even (if you've got the disk space—and patience—for *high*-resolution files) to produce effects similar to those available on imagesetters with stochastic, or frequency-modulated, screening.

PICTURE PUBLISHER

Micrografx's Picture Publisher, for Windows machines, offers an incredibly impressive tool set, including a nicely implemented unsharp masking dialog box, and good control over halftoning. It's tonal correction is lacking, however, making it a less-than-perfect tool for working with scans.

TONAL CORRECTION

Picture Publisher is disappointing in the essential area of tonal correction. While it offers a nice dialog box for controlling gamma settings, there's no way to control input or output levels—at least not in an approachable manner. We haven't figured out any workarounds for this.

SHARPENING

Picture Publisher's unsharp masking controls (available via the Effects menu item) are very similar to Photoshop's, which we discussed in Chapter 21, *A Sharper Image*. The numbers in this dialog box work slightly differently, however.

The Radius setting works as follows: a setting of two yields blips on each side of a tonal shift that are one sample wide (two samples for the whole halo). A setting of three yields blips two samples wide (four samples for the whole halo). So to achieve halos 1⁄50 of an inch wide, as we suggest, divide the image resolution by 100, and add one.

(Image resolution ÷ 100) + 1

Use the resulting value for your Radius setting.

The Strength setting is equivalent to Photoshop's Amount setting, except that it doesn't provide the range. We recommend leaving Strength set to the default of 100 percent—and even that isn't really enough. To compensate for this, you might try reducing Radius by 50 percent of what's recommended above, then running the filter twice.

The Threshold setting works exactly like Photoshop's. Set it to three or four for most images.

CorelPhoto-Paint

CorelPhoto-Paint, which is bundled with CorelDraw, is remarkably similar to Picture Publisher. That's not surprising, since they were both developed by Z-Soft, of PC Paintbrush fame. We don't have much to say about Photo-Paint that we haven't already said about Picture Publisher, since the controls are essentially the same—only more limited. It looks like this is the same program, but an earlier version, with a different interface.

Photo-Paint has the same failings as Picture Publisher for tonal correction, and its unsharp masking options are severely limited. You can do some rudimentary image adjustment in this program, but if you're looking for quality work, look elsewhere.

CACHET

Billed as a "color editor," Cachet's main goal in life is preparing scanned images for print—correcting, adjusting, sharpening, and separating. It doesn't work with grayscale images, but it offers some tools and techniques unavailable in other programs.

The most important tools to know about in Cachet are its concatenated tonal adjustments, tone- and gamut-clipping alarms, and the sharpening feature.

CONCATENATED TONAL CORRECTION. The goal of avoiding data loss before all corrections are done informs Cachet's whole approach. Unlike Photoshop and most other programs—in which you use one adjustment tool after another, losing some data with each correction—Cachet doesn't alter the image until you've made all your changes. You actually make your adjustments to a low-resolution version of the file, for the sake of speed. Cachet then concatenates all those adjustments, and applies the result in a single pass (when you save or print the image).

TONE AND GAMUT CLIPPING. When you're adjusting an image in Cachet, you can click on the tone clipping alarm to get an idea of what parts of the image are being pushed to black and to white—in other words, where information is being lost. It's much like the tone clipping display in Photoshop, except that you have to click a button after each correction to see the clipping.

The gamut clipping alarm goes a step further, showing you which colors in an image are outside the range of the target output method. So you can bump up the saturation of an image a little bit at a time, for instance, clicking on the gamut alarm each time to see where colors are getting clipped.

SHARPENING. Cachet only sharpens an image when you save it or print it. The idea is that you can't change an image after it's sharpened (that's the last step in the process), so you shouldn't sharpen it until you're ready for output.

There isn't much control over sharpening in Cachet—just five levels, from low to maximum. In our experience the top two levels result in oversharpened, artificial-looking (even blurry-looking) images. Stick to the bottom three levels for best results.

DeBabelizer

DeBabelizer (currently only for the Macintosh) is a dream come true for those who want to perform a lot of conversions on bitmapped graphics files. At its most basic level, DeBabelizer can open a graphic file in any of a huge variety of formats, and save it in a different format. That in itself is useful, for instance if you have a graphic in GIF format (CompuServe's native graphic format) and want to convert it to TIFF for import into your page-layout program.

Beyond simple conversion, however, with DeBabelizer you can manipulate images using an astounding variety of methods. There are simple tools for editing an image—pencil-drawing, erasing, selecting, copying, and pasting. However, Photoshop-like tools are not where DeBabelizer shines. Rather, it's for overall adjustments to whole images or batches of images.

When it comes to remapping color palettes, for instance, you can convert between bit depths and color modes, adjust tonal values by altering their response curves, dither an image in color, grayscale, or black-and white, manipulate color channels individually, and a whole lot more.

You can even use Photoshop plug-in filters in DeBabelizer (though unfortunately, several of the Adobe-provided filters—such as Unsharp Mask—don't work).

AUTOMATED CONVERSION

You can build any of these functions into a script, and by executing the script, perform multiple actions on an image consecutively. You can also control DeBabelizer via AppleEvents scripting, but the internal DeBabelizer scripting interface makes scripts so easy to create, edit, and execute, there's not much need for AppleEvents.

Scripting automates multiple tasks, and DeBabelizer's batch-processing tools take that a step further. You can run multiple scripts on multiple files, completely automating what could be an extremely complex conversion and image-manipulation process.

Though there are a few things we'd like to see changed in DeBabelizer (you can't perform unsharp masking, and the interface for tonal adjustments isn't nearly as good as Photoshop's Levels dialog box), this is a great tool for graphics conversion and batch manipulation. The sheer quantity of options will be daunting for less-technical users (see Figure 27-7), but if you spend any significant amount of time dealing with file conversion or image adjustment, you should have DeBabelizer.

FIGURE 27-7
DeBabelizer's Open dialog box

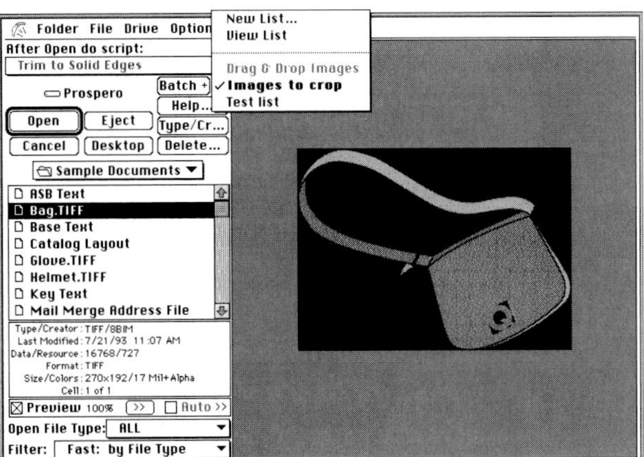

If this is just the Open dialog box . . . well, you can imagine.

Billions of Bits

When you're working with scanned images, you simply can't avoid big, bulky bitmaps. In order to handle those files properly, you often have to have industrial-strength image-manipulation tools. Without them, you could wake up late at night, sweat pouring down your neck after wading through pixelated landscapes and chatting with Mr. Sagan.

28 Page-Layout Applications

Putting It All Together

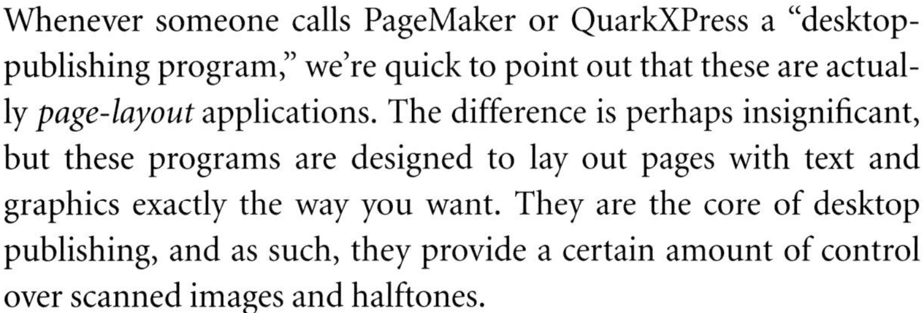

Whenever someone calls PageMaker or QuarkXPress a "desktop-publishing program," we're quick to point out that these are actually *page-layout* applications. The difference is perhaps insignificant, but these programs are designed to lay out pages with text and graphics exactly the way you want. They are the core of desktop publishing, and as such, they provide a certain amount of control over scanned images and halftones.

In this chapter we want to discuss briefly the major issues of halftoning and scanning with PageMaker and QuarkXPress.

PageMaker

PageMaker, the granddaddy of desktop publishing programs, has taken a big leap forward in its control of screens with version 5.0. Its print dialog boxes, support for PostScript Printer Description (PPD) files, and scan-related tools give you many of the tools you need to import scans and print them (and everything else) out the way you want them.

Controlling Screens

The main control for screen frequency and angle in PageMaker is in the Print Color dialog box (choose Print from the File menu, then click the Color button; see Figure 28-1). To set the screen frequency and angle, select Custom from the Optimized screen pop-up menu, and type the frequency and angle values you want in the appropriate fields (the default values for these fields are defined in the PPD file for the printer you have selected in the Print Document dialog box). If you choose Default from the popup, you get the output device's default screen settings. (See Chapter 13, *Controlling Halftone Screens*, for information about defaults.)

FIGURE 28-1
PageMaker's Print Color dialog box

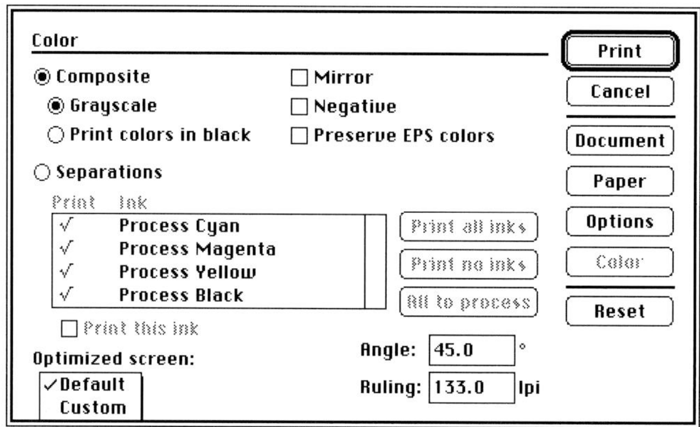

COLOR SCREEN FREQUENCIES AND ANGLES. If you're printing color separations to an imagesetter, the screening controls change somewhat. The optimized screen pop-up presents you with a list of choices for different screen frequencies and resolutions (see Figure 28-2).

Select one of the optimized screens, then click on the different process colors in the Separations box. Notice that the angles and rulings are different for different colors, presenting options that are (presumably) the best choices for that screen frequency. Notice also that the frequencies aren't exactly the same as what's specified in the Optimized screen pop-up.

As we mentioned in Chapter 12, *Frequencies, Angles, and Moirés*, it's often necessary to adjust screen frequencies for the four process colors to avoid moiré patterns. These are the best settings for a given frequency/resolution combination, as determined by Adobe and the imagesetter vendor. You can change the screen values for

FIGURE 28-2
Optimized screens for the Linotronic 330 with RIP 30

198 lpi / 3386 dpi
101 lpi / 2540 dpi
132 lpi / 2540 dpi
157 lpi / 2540 dpi
112 lpi / 2032 dpi
89 lpi / 1693 dpi
132 lpi / 1693 dpi
✓101 lpi / 1270 dpi
109 lpi / 1270 dpi
104 lpi / 3386 dpi / HQS
120 lpi / 3386 dpi / HQS
133 lpi / 3386 dpi / HQS
199 lpi / 3386 dpi / HQS
100 lpi / 2540 dpi / HQS
112 lpi / 2540 dpi / HQS
120 lpi / 2540 dpi / HQS
138 lpi / 2540 dpi / HQS
150 lpi / 2540 dpi / HQS
96 lpi / 2032 dpi / HQS
111 lpi / 2032 dpi / HQS
75 lpi / 1693 dpi / HQS
86 lpi / 1693 dpi / HQS
120 lpi / 1693 dpi / HQS
75 lpi / 1270 dpi / HQS
90 lpi / 1270 dpi / HQS
100 lpi / 1270 dpi / HQS
75 lpi / 846 dpi / HQS

each color plate individually, but we only recommend it for spot-color jobs. For process-color work, you're usually best off using the canned screen sets.

SCREEN SETTINGS FOR BITMAPS. You can specify the screen for individual grayscale bitmaps within a PageMaker pub via the Image Control dialog box (see Figure 28-3). Select a placed grayscale image, then choose Image Control from the Element menu. Click the Screened button, and type the frequency and angle you want in the appropriate fields. You can also choose between a round spot or line screen. These settings override the values in the Print dialog box, but only for the selected image.

FIGURE 28-3
PageMaker's Image Control dialog box

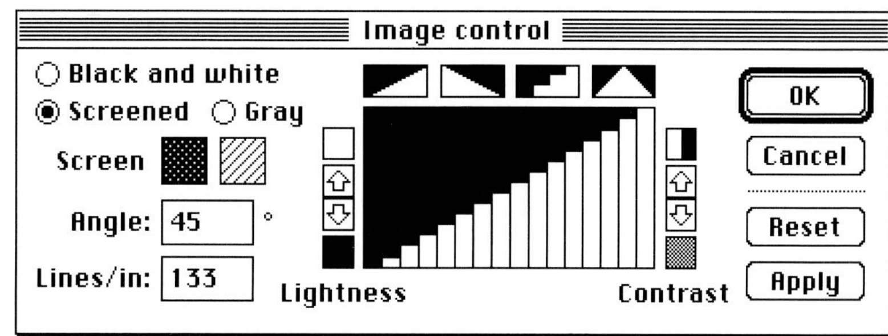

CONTROLLING SPOT SHAPE. Controlling the shape of your halftone spots from PageMaker is a little trickier; you need to edit the PPD for the printer or imagesetter you're using. Using a word processor or the like, open the appropriate PPD. (If you're using Microsoft Word, select Show All Documents from the pop-up menu and select the PPD. Then, if you're given a formatting choice, select Text.) You should see a section like the following.

```
*% Halftone Information ===============
*ScreenFreq: "133.0"
*ScreenAngle: "45.0"
*DefaultScreenProc: Dot
*ScreenProc Dot: "
{abs exch abs 2 copy add 1 gt {1 sub dup mul exch 1 sub dup mul add 1
sub }{dup mul exch dup mul add 1 exch sub }ifelse }
"
*End
*ScreenProc Line: "{ pop }"
*ScreenProc Ellipse: "{ dup 5 mul 8 div mul exch dup mul exch add sqrt 1
exch sub }"
```

There are three spots defined in this code: Dot, Line, and Ellipse. You choose which one is used with the "*DefaultScreenProc:" keyword. Just type in the name of the spot function ("ScreenProc") you want to use. You can add additional spot functions ("*ScreenProc Propeller," for instance), or alter the existing ones. For instance you might replace the code for "Dot" with code for a transforming elliptical spot, so images that have the round spot screen specified in Image Control actually print with elliptical spots. For more on halftone spots and PostScript spot functions, see Chapter 6, *The Glorious Spot*, and Chapter 26, *Spot Functions.*

BITMAP PRINTING OPTIONS

In the Print Options dialog box, you'll find four options for printing bitmaps: Normal, Optimized, Low TIFF Resolution, and Omit TIFFs.

NORMAL. This prints all the data in placed bitmaps—the full resolution of the source file.

OPTIMIZED. This only prints as much data as is needed for the specified screen frequency—a resolution of two times frequency—downsampling the bitmap at print time. While we generally prefer to set up our resolution properly in the first place when we scan and place a bitmap, this option is great when you're creating a document for high-res, high-frequency output, but proofing on a laser printer. Choose Optimized for your proof output, and PageMaker only sends as much information as is necessary, greatly reducing transmission and print times. Switch back to Normal for final output. And yes, Optimized does pay attention to special screen settings you've applied via Image Control.

LOW TIFF RESOLUTION. This option just prints the low-resolution screen rendition of a TIFF image. It's good for rough proofs.

OMIT TIFFS. This is when you want to go to the extreme. Selecting this option makes PageMaker print a box with an X in it instead of the TIFF. It's great when all you care about is proofing the text and general layout. It's also used to omit image files when you're working with OPI, and the low-res TIFF preview is replaced by a high-resolution image.

IMAGE CONTROL

We mentioned above that you can control screens for placed bitmaps in Image Control. You can also adjust their tonal values by moving the 16 bars in the dialog box. However, we find this tonal adjustment method almost useless except for simple things like screening back black-and-white clip art. You can't see a histogram of the image, or control the shadow and highlight points explicitly. And since sharpening should occur *after* tonal correction, and PageMaker doesn't provide sharpening, the tonal correction isn't much use. If you need image control, use your scanning or image manipulation software before placing the image in PageMaker.

COMPRESSING TIFFS

PageMaker can create compressed versions of TIFF files without leaving the program. Choose Place, select a TIFF file in the Place dialog box, and hold down the keys specified in Table 28-1 while clicking OK. PageMaker creates a new version of the file, in compressed form, with the filename extension specified in the table. For this to save any disk space, of course, you have to then go back and delete the uncompressed file.

TABLE 28-1
Compressing TIFFs with PageMaker

In the Place dialog box, select the TIFF to compress/decompress, and press these keys while clicking the OK button.

TIFF compression	*Keys to hold down*	*Resulting file extension*
Medium	Command-Option *or* Control-Alt	P
Maximum	Add Shift	L, LD, or LD2
Decompress	Command or Control	U

You have your choice of medium or maximum compression. Medium compression is faster to compress and decompress, but the files are larger than with maximum compression.

QUARKXPRESS

We've worked with QuarkXPress since the very beginning (this is evident if you've seen *The QuarkXPress Book*), and have seen it transform from a minor player in the DTP arena to a major leaguer on everyone's to-buy list. Curiously enough, over the past six or seven years, very little having to do with scanning and halftones has changed, at least on the surface. Here's the lowdown on these issues within QuarkXPress.

CONTROLLING SCREENS

There are three areas in which you can control QuarkXPress's halftone screening; two are in the Page Setup dialog box, and one is in the Edit Color dialog box.

HALFTONE SCREEN. One of the last items in the Page Setup dialog box is actually one of the most important: the Halftone Screen field (see Figure 28-4). The screen frequency that you enter here is applied to all TIFF and PICT images, as well as all tint values throughout your document (like tinted type, rules, background, and so on). You can override this value for TIFF images by using the Other Screen feature (see "Screening Control," next).

USE PDF SCREEN VALUES. When you turn the Use PDF Screen Values checkbox on, QuarkXPress looks at its internal PDFs (printer description files), as well as any that you've added from your printer manufacturer, for halftone screen angles and frequencies that are close to the one you typed in the Halftone Screen field. For example, if you are printing to a Linotronic at 133 lpi and you request a 45-degree angle, the magenta plate will print at 108.503 lpi and 19.983 degrees.

These values have been determined by the printer manufacturer to be the best process-color angles for the given screen frequency and resolution. Note that they might be far off from what you requested, but in theory they'll reduce moiré patterns. If you're printing to an imagesetter that uses Balanced Screening, HQS, or another high-quality screening system, you should turn this checkbox off (but check with your service bureau first). Turning the

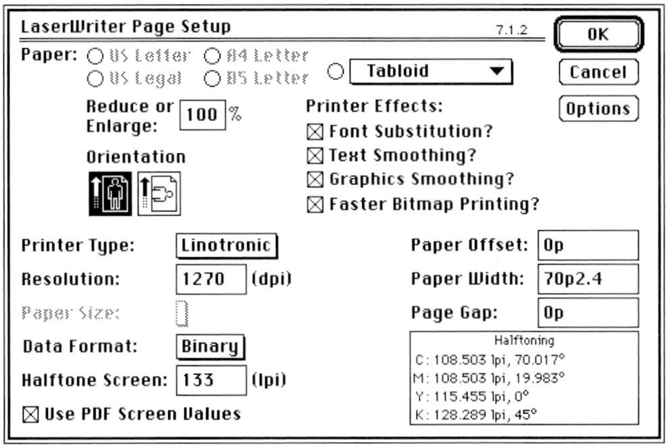

FIGURE 28-4
QuarkXPress's Page Setup dialog box

checkbox off reverts the halftone screens to the traditional values, and leaves it up to the imagesetter's screening filters to give you the proper frequency and angle (see Chapter 11, *Rosettes and Moirés*).

SPOT COLORS. If you're printing tints of spot colors alongside or on top of process colors, then you'll need to pay careful attention to the Screen Values pop-up menu in the Edit Color dialog box. QuarkXPress doesn't let you type in your own screen frequency and angle for the spot color plate, but it does let you assign the screen values from a process color. So if you want a green spot color tint to overprint on a black tint, you want to set Green's Screen Value pop-up menu to "Yellow" or "Magenta" or anything but "Black" (which is what it defaults to). This is far from the best of all possible worlds, but it's a step in the right direction, and can get most people out of trouble most of the time.

SCREENING CONTROL

As we noted earlier, you can always override the screens of grayscale TIFF and bitmap-only PICT images from within QuarkXPress by selecting the picture box with the Content tool and selecting Other Screen from the Style menu (or just type Command-Shift-S; see Figure 28-5).

FIGURE 28-5
Picture Screening Specifications dialog box

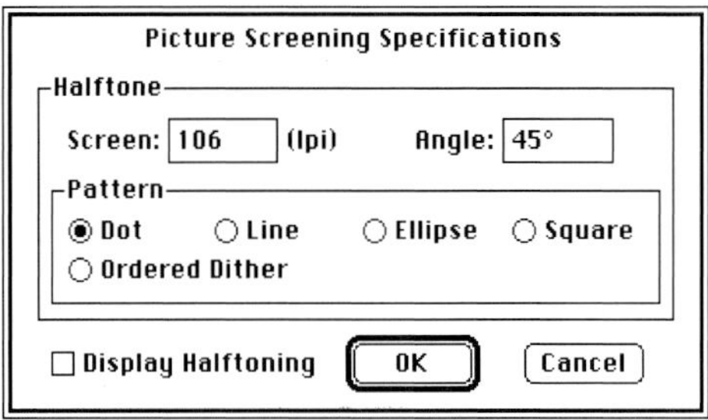

While everything in this dialog box is self-explanatory, you should especially note the Display Halftoning feature. This lets you actually see (vaguely) what an image will look like at the halftone settings you've specified. Of course, this is only really appropriate for low-frequency special effects (say, under 40 lpi).

PICTURE CONTRAST

Although QuarkXPress's ability to adjust tonal values in TIFF images is better than PageMaker's (see "Image Control," earlier in this chapter), it's still pretty pathetic when it comes to doing any real work. It does offer some good features such as posterization and pencil tools for manipulating the contrast curve (see Figure 28-6), but because you can't see a histogram or control any levels with precision or sharpen an image after correcting it, we just can't recommend you do much with it other than screening back black-and-white artwork. Instead, use Photoshop or another image manipulation program to prepare your images, then import them into your QuarkXPress documents.

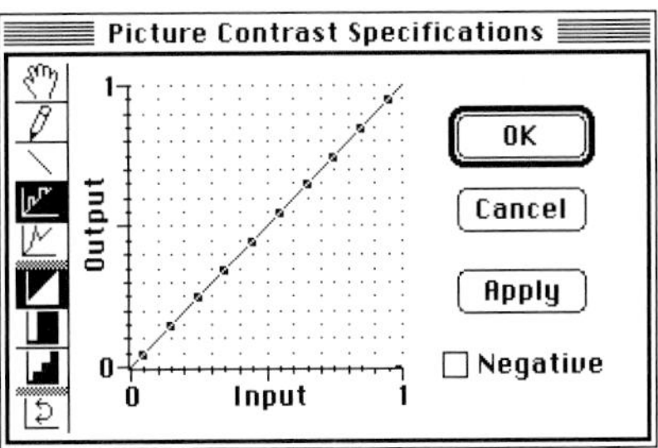

FIGURE 28-6 Other Contrast dialog box

DOWNSAMPLING

Back in Chapter 19, *Choosing Resolution*, we said that you never, ever needed to have an image resolution above two times the screen frequency. QuarkXPress enforces this in a way by trying not to send more than that resolution to the printer. So, if you have a 300-spi grayscale image, and you're printing at 53 lpi (the standard for most desktop laser printers), QuarkXPress only sends 106 dpi of information. This is called downsampling, and because it never cuts off less than what you need, you'll almost never have a problem.

QuarkXPress also downsamples line-art TIFF images to the resolution of the printer itself. For example, if you print a 800-spi image to a 300-dpi laser printer, QuarkXPress only sends a 300-spi image. This saves a great deal of time, and sometimes money, too.

Note that downsampling only works for TIFF and PICT images. QuarkXPress has no control over what does or does not get sent in EPS images.

As we write this (Fall 1993), there is an argument about how best to turn this downsampling off. The only way to do this now is by using the free Bobzilla XTension's Full Resolution Output feature (under the Item menu). However, this only works for one image at a time and is kind of a hassle. With any luck, in the future there will be a more elegant solution.

Why would you want to turn downsampling off? Because most applications' downsampling is pretty dumb. If you import a 200-spi image and scale it down to 25 percent, the final image is 800 spi. When you print that, QuarkXPress only sends a fraction of that to the imagesetter, right? Well, the fraction that it sends might not be the best possible choice and the image may appear mottled. The answer is really just to make sure your image is the proper resolution and size before you bring it into QuarkXPress to begin with.

Putting It Together

Page layout is where it all comes together: the scans, the text, the illustrations, the design. Without a solid understanding of how your page-layout program handles scans and halftones, there's no telling what could happen in your final artwork.

By the way, there are obviously other page-layout applications on the market, including PublishIt Easy and FrameMaker. We're covering only PageMaker and QuarkXPress here because they are the leaders in the graphic arts market and they provide the most control over halftoned images. As much as we hate to say this, we can only suggest that you check your software's manual for details on how your program handles halftones.

In the next chapter we take a final look at application-specific issues when we tackle the most commonly-used illustration packages.

29 Illustration Applications

Drawing Pictures

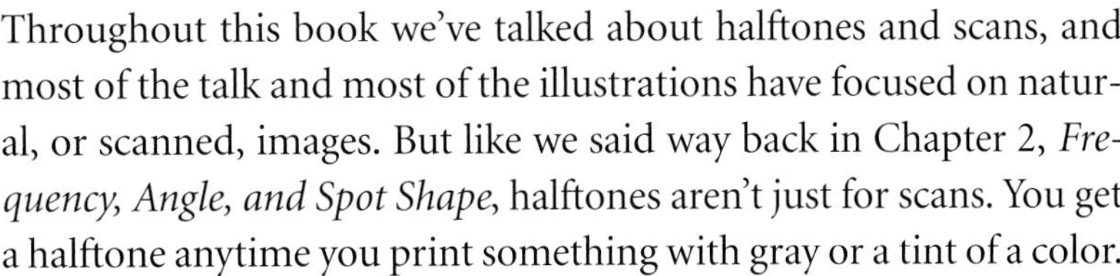

Throughout this book we've talked about halftones and scans, and most of the talk and most of the illustrations have focused on natural, or scanned, images. But like we said way back in Chapter 2, *Frequency, Angle, and Spot Shape*, halftones aren't just for scans. You get a halftone anytime you print something with gray or a tint of a color.

So in this last chapter we want to focus on illustration programs that create primarily synthetic, object-oriented artwork such as lines, boxes, circles, text, and so on. Although there are many of these programs, the primary ones (read: the ones you probably have on your computer) are Adobe Illustrator, Aldus FreeHand, and CorelDraw.

Adobe Illustrator

Everything we could say about scanning and halftones in Illustrator could fit on the head of a pin. That's because Illustrator offers no controls in these areas. But here's the rundown nonetheless.

Halftone screens. You can't specify screen frequency and angle in any way. If you print directly from Illustrator, you get the device's halftone default. We can't for the life of us figure out why they never added this feature.

EPS FILES. You can generate EPS images from Illustrator, but there's no screening information built into those, either. All screening is controlled by the application in which you place the EPS file, or by the device defaults.

PLACING IMAGES. Pictures that you place on an Illustrator page must be EPS files. Period. If there's screening information in the EPS image, it is applied only to that picture; the rest of the Ilustrator artwork is unaffected.

ADOBE SEPARATOR. The only way to control screening of Illustrator files is to open them with another program and control the screening there. One program that can do that is Adobe Separator, which is bundled with Illustrator. This program lets you set the screen frequencies and angles for each color in the Illustrator file (see Figure 29-1). Actually, Separator works with almost any PostScript file; it doesn't have to be from Illustrator. Of course, you don't need to use Separator to print Illustrator files. If you save them as EPS files, you can just as easily place and print them with PageMaker, QuarkXPress, or a variety of other programs.

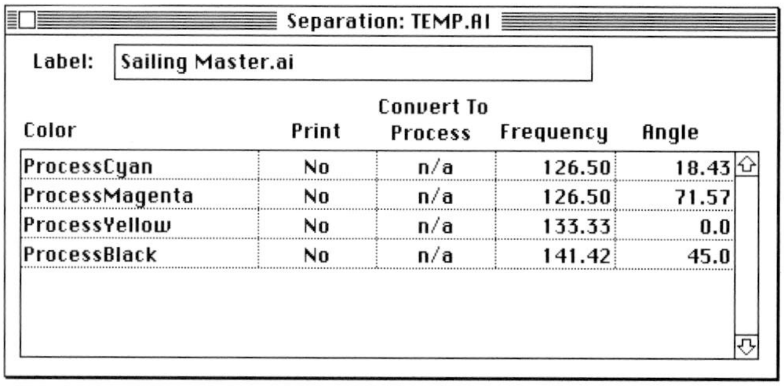

FIGURE 29-1
Adobe Separator dialog box

Separation: TEMP.AI

Label: Sailing Master.ai

Color	Print	Convert To Process	Frequency	Angle
ProcessCyan	No	n/a	126.50	18.43
ProcessMagenta	No	n/a	126.50	71.57
ProcessYellow	No	n/a	133.33	0.0
ProcessBlack	No	n/a	141.42	45.0

ALDUS FREEHAND

When it comes to working with halftone screens, Aldus FreeHand really gives you room to breathe. FreeHand lets you control screens when printing, control screens for individual objects within illustrations, and do some minor tonal adjustment on grayscale images. Let's look at each of these a little more closely.

CONTROLLING SCREENS

There are two places in FreeHand where you can control screens—in the Print Options dialog box (this applies to the whole print job), and in the Halftone Screen dialog box (this applies only to selected objects).

PRINT OPTIONS. To get to the Print Options dialog box (see Figure 29-2), choose Print, then click the Change button (no, *not* the Options button). Select a printer type, and FreeHand looks in the PostScript Printer Description (PPD) file for that printer. The PPD files provide—among other things—a list of frequencies that appear in the dialog box's Optimized Screen popup menu. You can select one, or choose Custom to type in your own screen frequency.

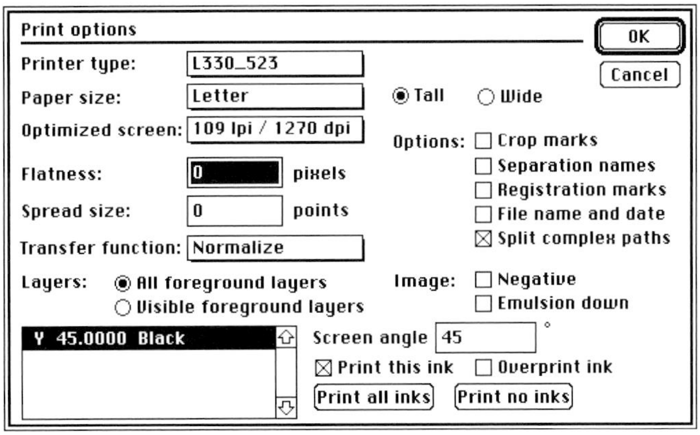

FIGURE 29-2 FreeHand's Print Options dialog box

You can control the halftone screen angle for a particular color at the bottom of the dialog box. Select the color, and if you're using an optimized screen set, the Screen Angle field shows you the defined angle for that set. You can alter the angle for any individual color (primarily useful for spot-color jobs). You can't alter frequency for individual colors, however.

Note that the screens you specify in the Print Options dialog box only apply when you're printing from FreeHand. They have no effect on exported EPS files.

HALFTONE SCREEN DIALOG BOX. To specify a special halftone screen for an individual element or group of elements within a FreeHand page, select the element(s) and choose Halftone Screen from the Attributes menu (see Figure 29-3). Here you can specify screen angle, screen frequency, and spot shape for the selected objects.

FIGURE 29-3
FreeHand's Halftone Screen dialog box

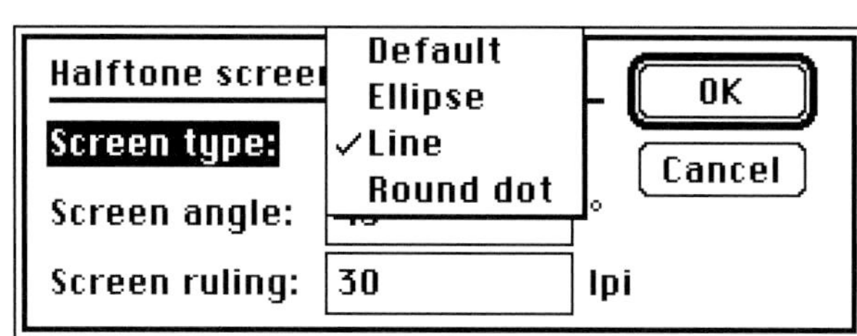

As we said back in Chapter 13, *Controlling Halftone Screens*, the object-specific values you set in the Halftone Screen dialog box override the application settings in the Print Options dialog box. These settings are also included when you export the file in EPS format.

This is just what you want if you're trying to create special effects for images, but beware if you apply it to a color object. When you specifically set the screen for an object, all the separations for that object print with the same angle and frequency—the one you specified. Document-level frequency/angle settings for individual colors are ignored for these objects.

You can control the spot shape for your FreeHand documents, however, without messing around with the frequency and angle. Just select everything on the page, and choose the spot shape you want in the Halftone Screen dialog box. Don't type anything in the Screen Angle or Screen Ruling fields. The spot shape you've chosen will apply to all the items in the document (even when you save it as an EPS), but frequency and angle aren't affected. They're still controlled by the printing application.

To alter the spot shapes in FreeHand's PPD files, see the section on PageMaker in Chapter 28, *Page Layout Applications*. Or even better, see Chapter 7, *Printing*, of Olav Kvern's book, *Real World FreeHand*.

TONAL ADJUSTMENTS

FreeHand also lets you make tonal adjustments of grayscale and black-and-white bitmaps via the Image dialog box. Just select a bit-mapped image, and choose Element Info from the Element menu (see Figure 29-4).

FIGURE 29-4
FreeHand's Image dialog box

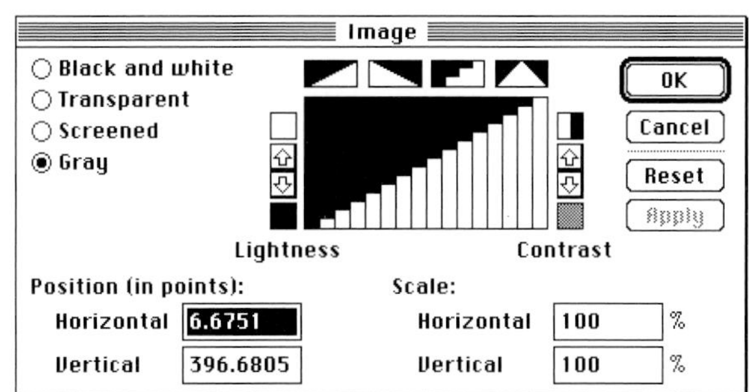

Curiously enough, this dialog box looks remarkably similar to PageMaker's Image Control dialog box. The tonal adjustments—16 adjustable level controls in all—are the same as PageMaker, and just about as useful (in other words, hardly at all). You might want to use them to screen back black-and-white images to gray, and other simple adjustments like that. However, if you want real tonal adjustment for photographic images, look to your scanning or image-editing software (like Photoshop) instead.

The Black-and-White and Gray buttons in FreeHand work the same as their PageMaker counterparts. You can use them to make a grayscale image print in black-and-white (with a 50-percent threshold: everything under a 50-percent tint goes to white, and everything above it goes to black), or a black-and-white image print in grayscale (so you can tint it). The Transparent button works just like the Black-and-White button, except it makes white sample points transparent; objects in back are visible behind them.

Note that the Screened button has no real effect, except on screen display; to control the screening of grayscale images in FreeHand, you use the Halftone Screen dialog box, as discussed above.

CorelDraw

CorelDraw provides all the essential controls for handling halftone screens with especially impressive object-level controls (including a mind-bending array of built-in, special-effect spot functions).

Imported Bitmaps

You can import bitmaps into CorelDraw in a variety of formats, but you can't do much with them aside from rotating and scaling them. You can, however, apply colors to bilevel bitmaps. Your best bet is to use an image-manipulation program for tonal correction and sharpening before importing images.

Controlling Screens

CorelDraw's Print Options dialog box is the main area for controlling halftones at a document-wide level. You can type a screen frequency in the Screen field, and it applies to all subsequent print jobs (overriding the setting in the Advanced Printing Options dialog box). You can't include this setting when saving EPS files, however (see Object-Level Screening Controls, below).

Process color. You can control the frequency and angle of each process-color ink from the Print Options dialog box. First, select Print as Separations, then click on the Separations button. In the Separations dialog box that appears, select Use Custom Halftone to change the angles and frequency for each process color. The default values—the ones that appear unless you've changed them—depend on the resolution of the printer you've selected.

Instead of using PPD files to store default frequencies and angles, CorelDraw uses its own information from the CORELPRN.INI file. The file has a section for each printer resolution that includes sets of frequencies and angles. Each resolution looks like the following.

```
[1200dpi]
Default=121lpi
85lpi=71.5651,18.4349,0,45,94.8683,94.8683,90,84.8528
90lpi=71.5651,18.4349,0,45,94.8683,94.8683,90,84.8528
106lpi=70.0169,19.9831,0,45,102.523,102.523,109.091,121.218
110lpi=70.0169,19.9831,0,45,102.523,102.523,109.091,121.218
121lpi=70.0169,19.9831,0,45,102.523,102.523,109.091,121.218
```

The resolutions included in the CORELPRN.INI file are: 300; 600; 635; 846; 900; 1,000; 1,016; 1,200; 1,219; 1,270; 1,500; 1,524; 1,693; 1,800; 2,032; 2,400; 2,483; 2,540; 3,048; 3,251; and 3386. Plus, because this is simply a text file, you can edit all the resolutions, frequencies and angles, if you choose.

SPOT COLOR. There's no way to control frequency or angle for spot colors in CorelDraw, except on an object-by-object basis via the Object Fill dialog box, which we discuss next.

OBJECT-LEVEL SCREENING CONTROLS

You can specify a spot function, frequency, and angle at the same time as you select line or fill colors in the pop-out menus for individual objects—but only for spot (read: Pantone) colors, including black. These settings not only affect the objects when printing from CorelDraw; they're also included in saved EPS files. As we said above, Corel offers a pretty impressive list of spot functions.

DRAW, PARDNER

Two hundred years ago no one used halftones. When they wanted to print a picture with grays or tints, they had someone etch it into wood or metal. Crosshatching was an art form, and the fine line art they created was so good that we still find it used today. In the subsequent years, the technology has progressed by leaps and bounds, but the concepts have held fast. Halftones—whether in illustrations or photographs—are subtle and can be beautiful. It's all in the hands of the creator: you.

We hope you've enjoyed this waltz through the process of creating scans and halftones. We have no doubt that with all this information under your belt, in no time you'll be roaring through the desktop-publishing jungles, scanner in one hand, imagesetter in the other—a true DTP professional.

Index

Permissions and Copyrights

Photographs of the Earth and moon used on front and back covers, color Page C, and the part, chapter, and other opening pages of the book, courtesy National Aeronautics and Space Administration.

The following images are reprinted with permission from the Special Collections Division, University of Washington Libraries.

Figures 2-2, 2-4, 10-4, 15-1, pages 10, 11, 79, 108, Photo #659
Figures 3-1, 20-4, 20-5, 20-6, pages 16, 160, 161, 162, Photo #80.A.W&S
Figure 4-1, page 24, Photo by A. Curtis, Negative #30744.
Figures 6-2 through 6-9, pages 41, 43, 44, 45, 46, 47, 49, 50, Photo by Farquharson, Negative #11.
Figure 8-3, page 60, Negative #NA3731
Figure 21-1, page 172, Photo by Goetzman Photo
Figure 21-6, page 176, Photo by Clifford Photo, Negative #UW9685
Figure 21-7, page 177, Photo by Todd, Negative #UW neg 10511
Figure 21-9, pages 181–184, Photo by Lee, Negative #20056

Authors' photo on back cover courtesy John Cornicello.

Illustrations used as Figures 1-2, 2-1, and 2-3, pages 5, 10, and 11, © 1993 Steve Stankiewicz.

Illustration used as Figure 16-4, page 12, courtesy Simon Tuckett.

Udaipur in Figure 20-2, pages 156–157; and Lake Palace Hotel, India, on Color page C, courtesy of Carol Thuman.

Bust of Pericles, Figure 20-11, page 167; Rick's graduation, Figure 24-2, page 206; S Kimble, Figure 24-5, page 209; stained glass in French church, and Charles greeting Koons family on Color page B, courtesy Glenn Fleishman.

Figure 21-5, page 175; Cactus, Jellybeans, Tulips and Breakfast on Color pages A, B, C, and D courtesy PhotoDisc.

Give Us a Holler

As we said way back at the beginning of the book, it took us over five years to finally get around to writing this book. In that time, it seems like the whole world of scanning and halftones from the desktop has change a thousand percent. And there's no sign that the changes will slow down in the future. One way that we'll keep up with all of the new tools, techniques, and tribulations (so we can roll them into the second edition) is by keeping up with all of you. If you find techniques that you find helpful, if you've learned anything particularly good from this book, or if you find something that you think is just plain wrong, let us know. Here's how to get in touch.

David Blatner
Parallax Productions

Steve Roth
Open House

1619 Eighth Avenue North, Seattle, WA 98109

David's e-mail
CompuServe: 72647,3302
America Online: parallax1
AppleLink: parallax1

Steve's e-mail
CompuServe: 72260,3100
America Online: sroth

Colophon

How We Made This Book

We wrote this book on David's Macintosh IIsi and PowerBook 170, and Steve's Centris 650 using Microsoft Word. We applied paragraph styles in the Word documents. The index was compiled painstakingly by hand.

Type. Body copy and running heads are set in the Adobe Minion family; headings and other display type are set in Copperplate 33bc.

Illustration, design, and production. We manipulated and created illustrations using Adobe Illustrator 5, Aldus FreeHand 3.1, Adobe Photoshop 2.5, and LaserTalk 1.3. The book was designed and produced by Glenn Fleishman of Open House in QuarkXPress 3.2 on a Macintosh Duo 210 with a Duo MiniDock. We proofed pages on a LaserWriter II NTX, Ricoh PC Laser 6000/PS, and LaserMaster Unity 1000.

Images. The images of the Earth on the front and back covers were scanned on an Optronics ColorGetter Plus drum scanner at Seattle Imagesetting from a 35mm transparency. The photograph of the authors on the back cover was shot on 35mm color reversal film, and scanned directly to Photo CD. Most of the images throughout the book were scanned on a Hewlett-Packard ScanJet IIc flatbed scanner. We grabbed screen shots with Exposure Pro and Tiffany.

Output. The book was imageset on an Agfa SelectSet 7000 at 2,400 dpi. The screen frequency is 133 lpi throughout, except where noted. All pages were output as imposed half-flats at Seattle Imagesetting by Neil Kvern using Aldus PressWise. The cover was designed and produced by Lee Sylvester and output on the SelectSet.

Place
29¢ stamp
here

Peachpit Press
2414 Sixth Street
Berkeley, CA 94710

Pixel Ink's
Interactive Resolution Tutorial Disks

The Pixel Ink's Resolution Tutorial explains everything you always wanted to know about resolution, and more. It shows, through a highly interactive process, how resolution can affect your image quality, blends and file sizes. The disk contains over six megabytes of colorful graphics, sound, animation and interactive calculations packed onto two disks, and comes with a four-color booklet with samples of images and blends.

Available for Macintosh only. Requires a 13-inch or larger color monitor, System 6.05 or later, and eight megabytes of RAM. Comes on high-density disks.

Yes! I want the **Resolution Interactive Tutorial** disks for **$32 US + $4 s/h.**
Macintosh high-density disks . . . . $ **36**
(California residents, please enclose $38.56)

METHOD OF PAYMENT

☐ Check or money order
☐ Credit Card ○ Visa ○ MasterCard ○ American Express

Card No. ______________________ Expires __________

Signature ______________________

Phone ______________________

Name ______________________

Address ______________________

City, State ______________________

Zip or Postal Code ______________________ Country __________

Please fold card in half, enclose check or money order if any, tape sides closed, and attach 29¢ postage to reverse.